ON AESTHETICS IN SCIENCE

ON AESTHETICS IN SCIENCE

Edited
by
Judith
Wechsler

The MIT Press
Cambridge, Massachusetts, and London, England

This book was set in V-I-P Optima by Woodland Graphics, and printed and bound by The Murray Printing Company in the United States of America.

Library of Congress Cataloging in Publication Data
Main entry under title:
On aesthetics in science.
Bibliography: p.
Includes index.
1. Science—Aesthetics. I. Wechsler, Judith 1940–
Q175.0477 501 77–26175
ISBN 0–262–23088–7

to Ben

CONTENTS

CONTRIBUTORS

Judith Wechsler
Associate Professor, History of Art
Massachusetts Institute of Technology

Cyril Stanley Smith
Institute Professor Emeritus, Professor Emeritus of Metallurgy, Professor Emeritus of the History of Technology and Science
Massachusetts Institute of Technology

Philip Morrison
Institute Professor, Professor of Physics
Massachusetts Institute of Technology

Arthur I. Miller
Professor of Physics
University of Lowell

Seymour A. Papert
Cecil and Ida Green Professor of Education, Professor of Mathematics
Massachusetts Institute of Technology

Howard E. Gruber
Professor of Psychology
Director, Institute for Cognitive Studies
Rutgers University

Sir Geoffrey Vickers
Social theorist and author

PREFACE

The idea for this book developed from a course I taught at MIT between 1972 and 1975 on Aesthetics in Science and Technology. As an art historian at MIT, I wanted to encourage substantial links between the arts and humanities and the sciences.

Some students attended art history courses because in that setting they could raise questions of context and style, issues generally ignored in undergraduate science courses. Yet their images and analogies and their aesthetic criteria drew strongly on their science studies. I speculated that explicit recognition of aesthetic dimensions in scientific work might encourage a more historical and experiential understanding of science, while evincing certain shared concerns of art and science.

I therefore designed a course on aesthetics in science and technology focusing on the role of models and the process of modeling rather than the beauty of artifacts in nature and science. Many students appeared to view science as an immutable and absolute reality supreme in the hierarchy of disciplines. One of my initial goals was to encourage them to recognize scientific formulae, concepts, theories, and models as human creations affected by traditions, styles, and sensibilities. Models are made—not "there" to be discovered.

We studied "works of science" as one might works of art, examining the relation of form and content, the personal and social context in which a work was created, and the intention and application of the work. Developments in science were studied with regard to prevailing styles, schemata, and paradigms, referring to the theories of Gombrich and Kuhn.

Many students developed increasing awareness of the fit between their personal sensibilities and their chosen field. Their interests and inclinations were reflected in the subjects of their research papers or projects on an aesthetic question in their own field. Papers included "A Study of the Heart, Levels of Aesthetic Value in Form and Function: Teleology as a Factor in Aesthetic Quality" by a biology premedicine major; "A Mathematical Analysis of Frieze Pattern Symmetries by Early Cultures as an Index of Civilization" by a mathematics student with an interest in philosophy and the history of science; and a study of how to predict the structural and aesthetic properties of a new mix of construction materials by a student in architecture and engineering. Some physics students studied problems of visualization such as observations outside the visible range in astrophysics and the photographic means of a translation into digital form. Another report explored Einstein's aesthetic criteria in his early papers.

The visiting lecturers focused on an aesthetic dimension in their own fields, for example, the role of aesthetic judgment. I wanted to avoid generalizations on art and science, such as the more evident connections between the beauty of scientific artifacts and their relation to the aesthetic terminologies of art. The focus was to be on aesthetics in the processes of doing science.

While the impetus for the book grew out of the course, and four of the chapters were originally lectures, the book is not intended as a report of a course but as a further development of a set of ideas. Like the course, the book encourages awareness of the role of aesthetic judgment in science.

A grant from the National Endowment for the Humanities in 1975 provided support for preparation of the book. I wish to thank Victor Weisskopf and Gyorgy Kepes for their encouragement and advice in the planning stage of the course as well as to the faculty and students who participated over the years. Discussions with Charles Eames furthered my interest in and understanding of aesthetics in science. Jehane Burns of the Charles Eames office provided constructive critical comments and advice on the introduction. Judy Lebow was helpful in her transcriptions of the taped lectures and in locating bibliographical material. My husband, Benson Snyder, offered helpful suggestions and support throughout the preparation of the book.

Judith Wechsler

ON AESTHETICS IN SCIENCE

INTRODUCTION

Aesthetic sensibility plays the part of the delicate sieve.
Henri Poincaré

Scientists talking about their own work and that of other scientists use the terms "beauty," "elegance," and "economy" with the euphoria of praise more characteristically applied to painting, music, and poetry. Or there is the exclamation of recognition—the "Aha" that accompanies the discovery of a connection or an unexpected but utterly right realization in art and science. These are epithets of the sense of "fit"—of finding the most appropriate, evocative and correspondent expression for a reality heretofore unarticulated and unperceived, but strongly sensed and actively probed. The right formalism or model which "captures" this reality seems almost magical in its potency. Both art and science evoke the previously ineffable in making ideas and concepts clear, cogent, and manipulable.

Heisenberg recalls commenting to Einstein on the force of recognition he associates with aesthetic experience:

> You may object that by speaking of simplicity and beauty I am introducing aesthetic criteria of truth, and I frankly admit that I am strongly attracted by the simplicity and beauty of the mathematical schemes which nature presents us. You must have felt this too: the almost frightening simplicity and wholeness of the relationship, which nature suddenly spreads out before us. . . .[1]

Definition

But the role of aesthetic judgment is rarely mentioned in the corpus of science and mathematics. When scientists, however, reflect on their work, the development of concepts, and the theories that expound them, it is evident that intuition and aesthetics guide their sense of "this is how it has to be," their sense of rightness.

It is almost too obvious to say that if one believes science to have a singular and exclusive relation to reality and assume it to be synonymous with truth, then the idea of aesthetics in scientific judgment or cognition may seem capricious or marginal. One can still regard the products of science as beautiful (truth equals beauty). But if one views science as attempting to approximate reality, subject to experiment but not necessarily to verification, then there is latitude, and one can conceive that the choice of alternative hypotheses are subject to aesthetic factors. Karl

Popper's theory that "not the *verifiability* but the falsifiability of a system be taken as a criteria of demarcation," allows for that latitude. ". . . it must be possible for an empirical scientific system to be refuted by experience," he writes in *The Logic of Scientific Discovery* (1959).

The Oxford English Dictionary defines aesthetics as "things perceptible to the senses, things material" as opposed to "things thinkable or immaterial." Aesthetics by this definition might seem inapplicable to those areas of science whose operations appear to be purely intellectual-logical processes, mathematical formalisms.

Kant understood the limitations of such a definition when he commented in *The Critique of Pure Reason*, "concepts without factual content are empty; sense-data without concepts are blind. The senses cannot think. The understanding cannot see. By their union only can knowledge be produced."

In its dealings with "things perceptible to the senses," aesthetics comes to grips with *relations:* structure, context, schemata, similarity/dissimilarity, consonance/dissonance. (About isolated, simple sense impressions aesthetics has little to say.) These relations, which are sometimes separated from their "factual content" or perceptual substrate in science, are still subject to aesthetic preference, as they are in art or nature where this separation has not been made.

In this sense, Kant writes, aesthetics is "the science which treats the *conditions* of sensuous perception" (my emphasis).

A dichotomy exists in science between those like Bohr who assume that their starting point and base of verification is sense perception, and those, like Heisenberg, who believe that sense perception is an unnecessary limitation. Such a case is the contrast between Hertz and Mach: Hertz advocated a purely intellectual process in his "pure natural science," while Mach believed "every statement in physics has to state relations between observable quantities." Both pure science and science related to observation are subject to aesthetic judgment. The aesthetics of pure relations engages our minds as music does through harmonic relationships.

Intuition as well as aesthetic judgments operate in both approaches to science. In art, and in life, we acknowledge the place of aesthetics and intuition, but we don't readily associate these more tacit dimensions with the logical processes of science. Yet, as Norbert Wiener and Arturo Rosenblueth observe,

> An intuitive flair for what will turn out to be the most important general question gives a basis for selecting some of the significant among the indefinite number of trivial experiments which could be carried out at that stage. Quite vague and tacit generalizations thus influence the selection of data at the start.[2]

Scope

The contributors to this book all maintain that aesthetics is a crucial factor in the scientific process. Aesthetics is discussed in this collection not as a systematic discipline in philosophy, but as a mode of discrimination and response—a guideline for the *appropriateness* of a scientific expression (as in the chapters by Papert and Vickers).

Aesthetics is also associated with visualization (Miller), structure (Smith), metaphor, image, and analogy (Morrison, Gruber). The question is raised, How do aesthetic considerations affect the form, development, and efficacy of models?

Aesthetic sensibility also enters the appreciative mode. For the majority of practicing scientists, aesthetic criteria enter in the ways of response. The aesthetics of recognition is at work when we grasp an idea, understand how a principle operates, or how a solution was found. (This issue is discussed in all the essays.) Our admiration of Copernicus or Newton is of this order and can be likened to our aesthetic appreciation of Cézanne, Bach, or Milton. Science too can be a source of aesthetic delight.

As in art, aesthetics is subject to period styles as well as personal ones. It is bound to change in time, in both subject and locus. There is ample evidence to suggest changes in scientific style and taste, in the problems posed and the methods posited. While a sense of "fit" may be timeless, the context and connections in which a theory first emerges are affected by schemata and tradition. In the twentieth century, aesthetics as the appreciation of form has expanded to include process as well as product.

Process

The emphasis in this book is on process in science, the issue of modeling. The finished work, in science as in art, gives evidence of its process. However, the balance between product and process differ in art and science. In art, it is the finished painting, sonnet, or sonata which is normally the subject of our criticism or appreciation. Science doesn't exhibit products for aesthetic criticism in this way; there is hardly any recognized vocabulary of aesthetic criticism and response for science. The usual criterion for "success" of a scientific product—an equation, a physical model, or a written paper—is whether it works, that is, predicts, explains. Aesthetic judgments operate in the cognitive processes of arriving at that product.

Our contemporary interest in process has a history. At the end of the nineteenth century, developments occurred in the arts and sciences which challenged the belief in an objective reality unmediated by the subjective perceiver. In painting, the breakdown of "scientific" perspective meant an end to an established a priori way of translating reality onto a two-dimensional surface. Cézanne asserted subjective perception in the form of shifting perspectives. Similarly in literature, James Joyce composed his novels from the multiple view points of his characters.

It is in this period that science too began to recognize the individual, perceptual and intellectual screen in constructing models of reality, for example, in Poincaré's advocacy of intuition and Einstein's theory of relativity. These developments in the arts and sciences articulated the complex phenomenological relationship of subjective perceiver and objective reality.

Now more than ever we focus on the effect of process in art and science; our age is particularly self-conscious of cognitive modes. Because of Freud we are alert to latent as well as manifest content. The studies of Piaget, Bruner, and other cognitive psychologists indicate the role models play in our thinking. Systems analysis and conceptual art further prepare us to look at process rather than product.

Scientist's predilections are often revealed in their choices of imagery and metaphor. Niels Bohr, recalled Heisenberg, equates the language of poetry and physics:

> . . . when it comes to atoms, language can be used only as in poetry. The poet too is not nearly so concerned with describing facts as with creating images and establishing mental connections. . . . Quantum theory . . . provides us with a striking illustration of the fact that we can fully understand a connection though we can only speak of it in images and parables . . .[3]

Aesthetics viewed as a mode of scientific cognition may parallel a scientist's outlook and concern in other areas of life. Poincaré observed of mathematicians, "It is the the very nature of their mind which makes them logicians or intuitionalists, and they cannot lay it aside when they approach a new subject."[4] Werner Heisenberg and Niels Bohr often wrote of these links. Gerald Holton has examined the mapping of personal life onto science in Einstein's work.[5]

Intuition and Aesthetics

Bohr, Dirac, Einstein, Heisenberg, Poincaré, and others acknowledge intuitive and aesthetic judgment as decisive factors in the acceptance or rejection of a particular model.

Poincaré was convinced of the role of intuition in scientific process.

> Pure logic could never lead us to anything but tautologies; it could create nothing new; not from it alone can any science issue. In one sense these philosophers are right; to make arithmetic, as to make

geometry, or to make any science, something else than pure logic is necessary. To designate this something else we have no word other than *intuition*.[6]

Poincaré goes further, linking intuition with aesthetics:

It may appear surprising that sensibility should be introduced in connection with mathematical demonstrations, which, it would seem, can only interest the intellect. But not if we bear in mind the feeling of mathematical beauty, of the harmony of numbers and forms and of geometric elegance. It's a real aesthetic feeling that all mathematicians recognize, and this is truly sensibility . . . The useful combinations are precisely the most beautiful . . .[7]

Reinforcing Poincaré's trust of aesthetic judgment, Dirac commented on Schrödinger's not publishing his first version of the wave equation because it conflicted with empirical data:

I think there is a moral to this story, namely that it is more important to have beauty in one's equations than to have them fit experiment. . . . It seems that if one is working from the point of view of getting beauty in one's equations, and if one has really a sound insight, one is on a sure line of progress. If there is not complete agreement between the results of one's work and experiment, one should not allow oneself to be too discouraged, because the discrepancy may well be due to minor features that are not properly taken into account and that will get cleared up with further developments of the theory.[8]

Polarities

Dirac and Poincaré both appeal to aesthetic criteria yet the frame of their aesthetic differs. Poincaré's is a more geometric sensibility, Dirac's an abstract mode of deductive reasoning. But Dirac too is concerned with the beauty of form, of relativistic covariance or persistence of form. The capacity to visualize, Dirac maintains, does not advance new theories, but rather reflects a more basic need to picture and represent.

Many scientists have commented on the polarities, even the dialectic, of scientific imagination. Gerald Holton has described this phenomenon as the "thema and antithema" of science. Two basic forms of scientific thinking have been characterized as descriptive vs. structural science (Aristotle vs. Plato), "the concept of space vs. the concept of number" (Ernst Cassirer), pictorial models (process of operations) vs. functional models (functional mapping) (Wiener), logicians and analysts vs. intuitionalists and geometers (Poincaré).

Contrasting modes of scientific imagination is a recurrent issue in this book. Vickers distinguishes between aesthetic criteria and rational deduction in making judgments, Papert between intuition and pure logic in mathematics, Miller between visualization and nonvisualization in quantum theory, and Gruber between classic and romantic imagery in nineteenth century biology.

We know from the development of quantum theory that alternative explanations can coexist, be complementary: wave and particle, continuity and discontinuity, mathematical formalism and visualization, each of which "describes" the nature of atomic phenomena. (Both are needed because neither saves *all* the phenomena.)

It may be more difficult to specify the aesthetic properties of symbolic mathematical relations than those of models, metaphors, and images based on our experience and observation.

The aesthetics of pure mathematical relations and the strong emotion it evokes recall the formalist aesthetic in art—though at a more abstract level—since mathematical formulations sometimes eschew visualization. The principal criterion in this aesthetic was termed "significant form" by the art theorists Clive Bell and Roger Fry around 1914. Significant form implies the optimal unity and coherence of the compositional elements. Fry's position marked the beginning of the modernist aesthetic in art, with its lack of concern for representation and its focus on the beauty of formal relations. Significant form evokes a particular kind of reaction that Fry described as "aesthetic emotion," which, according to him, is more intense and focused than the ordinary emotion we experience in our daily lives. Mathematicians have said the same of their experience of pure mathematical relations.

Conclusion

The choice of orientation is not necessarily set in science by the problem but by a mode of thinking. Though there are constraints, there is no a priori essential epistemological way of seeing. Therefore the role of cognitive mode and aesthetic sensibility plays a vital part in the structure and style of the scientific process.

In summary, aesthetics is presented in this collection as a mode of cognition which focuses on forms and metaphors used in scientific conceptualizing and modeling. The attention to process in science and art leads to a consideration of the part played by paradigms (Kuhn) and personal style in discovery and invention.

Viewed as a way of knowing, aesthetics in science is concerned with the metaphorical and analogical relationship between reality and concepts, theories and models. The search in science for models that illuminate nature seems to parallel certain crucial processes in art, as Cyril Smith points out: they share a fundamental evocative quality.

The forms of symbolization are central to the study of aesthetics in science. Science uses a symbolic language, most commonly mathematics. New concepts necessarily expand the vocabulary and syntax of this language. The way new symbolic forms develop is in part influenced by aesthetic concerns.

Scientific ideas (models, hypotheses) develop within a cultural framework rooted in the scientist's time in history. There are cultural styles and traditions in scientific concepts. Though there has been a historically recurrent appreciation of elegance and economy, Victorian science, for example, often preferred more complicated formulations. An examination of aesthetic criteria reveals the pressure of value systems that are related to the scientist's social and cultural context. The ability to relinquish the predictability of classical physics and accept indeterminacy, probability, and complementarity, we associate with contemporary sensibility. So too the emphasis on process over product.

This book examines the aesthetics of formulae, theories, concepts, models, and processes. The aesthetic factors in cognition—manifest in both art and science, though until recently more recognizable in art—are continuous and broader than either. The essays by four scientists and two social scientists will hopefully enrich the reader's view of the nature of aesthetic cognition in the scientific process.

Notes

1. Werner Heisenberg, *Physics and Beyond* (New York: Harper & Row, 1971), p. 68.

2. Arturo Rosenblueth and Norbert Wiener, "Roles of Models in Science," *Philosophy of Science*, XX (1945): 317.

3. Heisenberg, *Physics and Beyond*, p. 210.

4. Henri Poincaré, *The Value of Science*, trans. G. B. Halstead (New York: Dover Publications, 1958), p. 15. (Originally published in 1905.)

5. Gerald Holton, "Mach, Einstein, and the Search for Reality," *Thematic Origins of Scientific Thought: Kepler to Einstein* (Cambridge: Harvard University Press, 1973).

6. Poincaré, *The Value of Science*, p. 19.

7. Henri Poincaré, *Science and Method*, trans. Maitland (New York: Dover Publications, n. d.), p. 59. (Originally published in 1908.)

8. P. A. M. Dirac, "The Evolution of the Physicist's Picture of Nature," *Scientific American*, May 1963, p. 47.

Cyril Smith's essay is concerned with the aesthetics of structural relationships and their metaphorical and analogical associations. He examines the ancient problem of parts and wholes. Smith observes, "In any physical or conceptual or social system, as one goes up or down the scale of magnitude, there is a continual alternation between the externally sensed qualities of unified "things" and their internal structure which changes in response to external exploratory contact . . . Styles in art, like the existence of identifiable phases in a chemical system, reside in the extension of a repeated local pattern of association of parts . . ."

J.W.

STRUCTURAL HIERARCHY IN SCIENCE, ART, AND HISTORY

CYRIL STANLEY SMITH

Introduction

The author is a metallurgist. When, years ago, I became more than casually interested in the history of my profession, I searched the written records of the last eight centuries without even approaching the beginnings of important techniques, but eventually I found that the earliest evidence of knowledge of the nature and behavior of metals was provided by objects in art museums. Slowly I came to see that this was not a coincidence but a consequence of the very nature of discovery, for discovery derives from aesthetically-motivated curiosity and is rarely a result of practical purposefulness.

Also, having spent many years seeking quantitative formulations of the structure of metals and trying to understand the ways in which the structures change with composition and with treatment, and the ways in which structure relates to useful properties, I have slowly come to realize that the analytical quantitative approach that I had been taught to regard as the only respectable one for a scientist is insufficient. Analytical atomism is beyond doubt an essential requisite for the understanding of things, and the achievements of the sciences during the last four centuries must rank with the greatest achievements of man at any time: yet, granting this, one still must acknowledge that the richest aspects of any large and complicated system arise in factors that cannot be measured easily, if at all. For these, the artist's approach, uncertain though it inevitably is, seems to find and convey more meaning. Some of the biological and engineering sciences are finding more and more inspiration from the arts.

It is important to note that the word "hierarchy" is used throughout this paper not in the the sense of a rigid boss/slave relationship, with control passing unidirectionally from top to bottom, but even more as the inverse, for it is the interlock of the smaller parts that generates the larger overarching structures. My model is not the command tree of figure 1 but is more like the

This article is an extension of notes made for an informal talk in the series on "Style" organized by the Cambridge Archaeological Seminar in the spring of 1974 and for one of the Danz lectures at the University of Washington in March 1975. The author dedicates this paper to the memory of Lancelot Law Whyte in grateful acknowledgment of stimulating conversations over a quarter of a century.

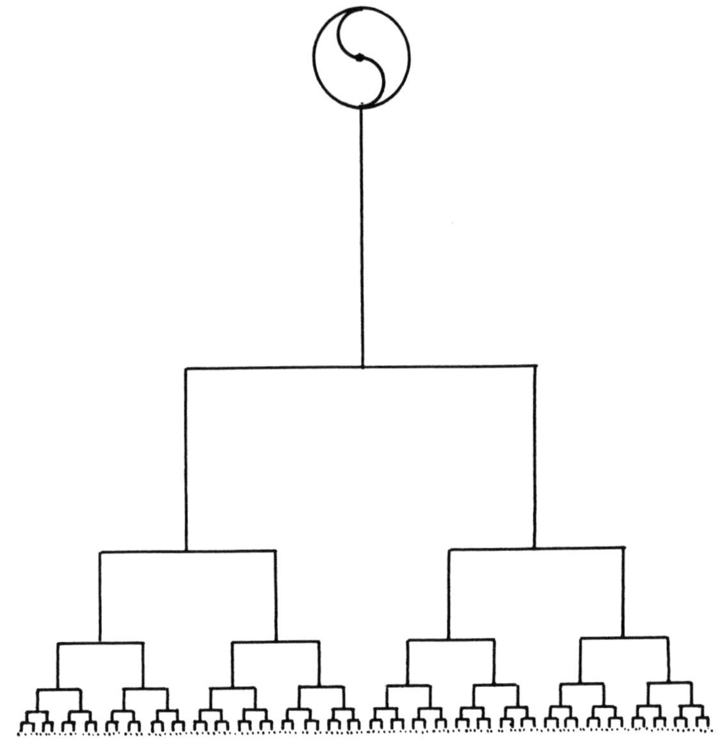

Figure 1. The tree of logic, representing the conventional view of hierarchy as a one-directional system of dominance. Compare with Figure 8 which shows an interlocking structural hierarchy.

self-forming, locally-diverse aggregate of figure 8. Though environment will strongly influence the behavior of small things within it, the environment itself comes from the interaction of individuals to produce ever-larger structural groupings. All things are interactive, both up and down the scale of clumpings. Though the larger groups necessarily incorporate the smaller ones, the aspects of hierarchy here considered are far more than mere inclusion, just as they are more than mere addition or mere dissection. One of my colleagues at MIT has pointed out that, since most complex structures arise without control from above, "anarchy" might be a better term than "hierarchy" for the subject of this paper. It would be good to avoid both terms for they are overloaded with political emotion, but I know of no better word than hierarchy to convey the idea of an interpenetrating sequence of structural levels.

The author's experience with the hierarchy of structural changes associated with the hardening of bronze, steel, and duralumin may seem like poor qualification for writing on the elusive subject of art, but I hope to show that it is not entirely irrelevant since style in a work of art is an aspect of its structure, and structures of all kinds have certain similarities regardless of what materials, objects, societies or concepts they relate to. The differences, and the problems, come in the identification of the significant units and the manner in which hierarchical entities successively present themselves as the size of the aggregate increases and as the scale of resolution and the character of what can be perceived changes with the method of observation.

The unique quality of a work of art depends on the manner in which its component parts are shaped and put together. It has style, but the style is unrecognizable except by comparison with other works having similar structure and overtones, an extension of an inevitable hierarchy. In chemistry, the characterization of a phase (whether crystalline or not) also depends on the way the parts are put together and the way in which this internal structure affects external contacts to build up materials on the scale at which we use them.

Structure provides a universal metaphor: the apparent mixing of metaphors throughout in this article is not entirely due to carelessness! *Everything* involves structural hierarchy; an alternation of external and internal, homogeneity and heterogeneity. Externally perceived quality (property) is dependent upon internal structure; *nothing* can be understood without looking not only at it in isolation on its own level but also at both its internal structure and the external relationships which simultaneously establish the larger structure and modify the smaller one. Most human misunderstanding arises less from differing points of view than from perceptions of different levels of significance. The world is a complex system and our understanding of it comes, in science, from the matching of model structures with the physical structure of matter and, in art, from a perceived relationship between its physical structure and the levels of sensual and imaginative perception that are possible within the structure of our brain's workings. All is pattern matching, with the misfits, if they can interact, forming a superstructure of their own as in moiré patterns or in beat notes.

As a rudimentary example, compare figures 2 and 3. The former represents the structural arrangement of atoms in an alloy such as a piece of brass, with two crystals of one phase with "atoms" in square array (the two being incompatible only because of their differing orientations), and a third crystal of a different structure based on triangles. More correctly, figure 2 is not the structure of brass but rather it is a sketch reflecting my idea of the structure, which itself already partially maps other conceptual and physical patterns and is rendered meaningful only by the possibility of establishing resonant matches with patterns in other minds.

Figure 3 is a detail of a recent etching by Tanaka Ryohei, who has skillfully used his needle and other devices to produce textural arrays in areas that define and depict walls, roofs, and other features that relate to each other to produce a meaningful picture on the scale of the whole (fig. 3a). In both figures 2 and 3 there is a just-visible detail in local pattern which the eye quickly passes over to appreciate the larger areas characterized by its extension. These areas in their turn are seen as units in a new pattern grouped within a larger composition. The units are essential, but it is not the units that signify, it is the pattern formed by the repetition of their relationship. Moreover, the distinction between one area and the next in either figure 2 or 3 comes from the incompatibility of their microstructures. The mismatch need not be geometric (at least not on a visible scale) but it may be based on any perceivable interaction in art or any physical interaction in matter—or, more commonly, on many different interactions superimposed. Seurat performed magic with small patches of color. In ceramic glazes the aesthetic units are sometimes identical with the areas of a physical phase.

From the viewpoint of a physicist, the hierarchy in figure 2 stops with the symmetry of the crystal lattices. The metallurgist sees, on a larger but still small scale, the two-phase microstructure in a sample of 60/40 brass, while the engineer sees a lump of brass, useful to be shaped into naval hardware, a plumbing fixture, a telescope body, or other object to play its role within the still larger structure of society.

A work of art has failed if it does not further enlarge in the viewer's mind to encompass and to relate many overtones and different levels of personal, emotional, social, and cultural themes

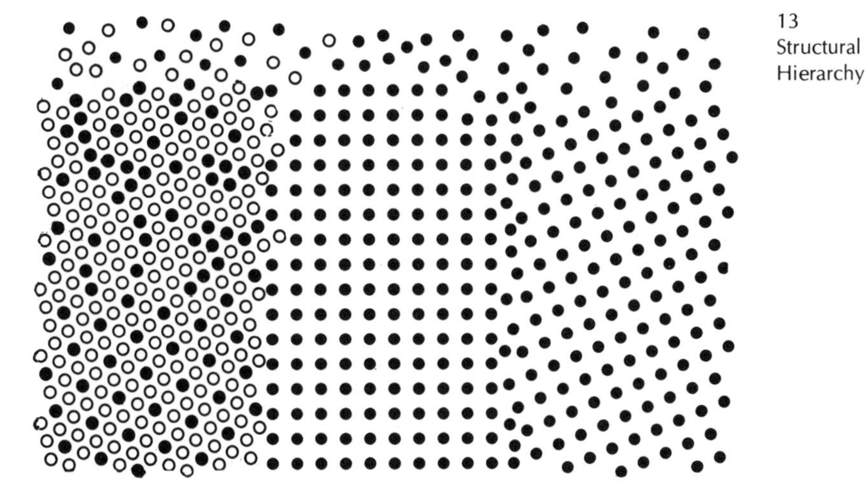

Figure 2. Diagram showing the arrangements of atoms in three cyrstals. Two are based on a square lattice and have a boundary between them resulting only from their different orientations, while the third crystal (on the left) is based on a triangular lattice and contains a ratio of three "white" atoms to each "black" atom, in regular ordered array at the bottom, disordered near the top. The disorder of the three or four atom layers at the top suggests the structure of a liquid.

Figure 3. Detail of an etching "House in Ohara" by Tanaka Ryohei (1975). Compare with Figure 2. The extension of different local patterns of juxtaposed fine detail is used to define areas representing different features in the overall design, Fig. 3a.

15
Structural Hierarchy

all joined together mentally, some for the first time and uniquely by the sensing of a larger pattern of harmony between imagination and understanding.

All things have internal structure. Externally, a thing may have form, and possess considerable grace in the balance of its parts, but it cannot be said to have a style unless some aspect of the relationship of its parts appears in other objects and by its replication provides a basis for a perceptive eye to group them together. Original creations must inspire copies. Style is the recognition of a quality shared among many things; the quality, however, lies in structure on a smaller scale than that of the things possessing the quality.

Style in art is closely analogous to the relationship between internal structure and externally measurable property that in science distinguishes one chemical phase from another. Similar problems of characterization, boundary identification, and mechanisms of change exist in both fields. Bulk properties of matter such as density, color, conductivity, crystal structure, or vapor pressure by which a chemical phase is identified are not a property of any of the parts (though they would not exist without them) but rather are external characteristics depending on the pattern of interaction between the atomic nuclei, electrons, and energy quanta and the extension of this pattern by repetition throughout the entire volume of the phase concerned. The pattern and the property both disappear when there are too few parts, and they change when an interface into another distinguishable phase is crossed. So with styles of art. They cannot be seen from inside, although the structure on which they depend can be. Some styles, like some phases (for example, liquid or solid solutions) can tolerate considerable diversity in the shape and constitution of their parts, while others (like simple molecules and covalent crystals depending upon the precise symmetry of nearest-neighbor interaction) are intolerant of substitution: if different parts are introduced the whole structure will adjust to a new form. Many important differences in the behavior of large systems arise from the differences in the degree of tolerance or intolerance of neighbors within the smallest groupings that compose them.

In the aesthetic case the identifying uniformity corresponding to the physical property test of a chemical phase is that of some psychological response triggered by the repetition of some detail of pattern or color relationship. Though equilibrium is not involved, there is a kind of phase rule relating the number of distinguishable types of features in a painting to the number of kinds of just-visible units with which the artist works and the manner and density of their packing. Etchers and engravers use cross hatching and stippling almost like crystal lattices to distinguish areas. In figure 3 the eye immediately senses an individual quality in each of the areas that, on close inspection, can be seen to arise from the array of lines that differ in width, uniformity, spacing, and orientation in a manner that is quite analogous to the way in which chemical phases are distinguished and given identity. Neither atoms nor lines are enough by themselves: It is their interplay that gives unity and character to the assembly even when the parts are invisible, and the assembly on one level—whether in matter or in the perception that underlies art—itself becomes a part in a larger aggregate with its own pattern of interplay.

Chinese landscape paintings, constructed as they usually are with a limited number of distinct types of brush stroke, show this particularly well. The literati painters of the Ming dynasty made use of this device very effectively, but it originated much earlier. Figure 4 is a portion of a landscape painting by the early Ch'ing painter Kung Hsien (ca. 1620–1698) and figure 4a a smaller detail. Each of the areas, for example, those suggesting trees, rocks, sky, or mountains, has an immediately sensed distinctive quality which relates it to others of the same kind within the painting as a whole, but which arises in the repetition within it of a particular type of brush stroke and the manner in which the brush strokes relate to their immediate neighbors. In a physical system the maximum number of phases that can coexist in equilibrium at a given temperature depend upon the number of atomic species present and how closely they are pressed together. Whether a boundary can exist between one area and another depends upon cooperative hierarchical interaction, on whether the valencies (desire for resonant communication) of individual atoms are compatible with those of the aggregate of their neighbors. Ultimately, the possibility of continuity or separation between two regions depends upon the numerical topological requirements of space filling—in whatever space and by whatever means of connection is appropriate for the occasion. The requirement of uniformity within the connected region of each "phase" is relaxed when equilibrium has not been attained, that is, when the system retains a memory of some previous structure. In these, and rather generally in works of art, gradients in microsymmetry exist on a scale comparable to that of the granularity. However, too much disorder can soon cause a phase to lose its identity, and a proliferation of gradients of color or texture in a painting quickly diminishes its visual impact. In systems of any kind effective diversity does not continue to increase with the complexity that accompanies a greatly increased number of units. Properties such as density, compressibility, or vapor pressure that are established by the interaction of a few atoms do not change much in larger aggregates, although others, such as strength and plasticity are enormously

18
Cyril Stanley
Smith

Figure 4. Landscape painting, "Wintry Mountains" by the early Ch'ing painter Kung Hsien (ca. 1620–1698). 165 × 49 cm. Collection of John M. Crawford, Jr. Note the use of brush strokes almost as atoms to form, to relate, and to distinguish the larger features. Fig. 4a. Detail of Fig. 4.

influenced by imperfections and larger clusterings. In visual perception the angular resolution limit means that the number of things that can be perceived at any one time remains approximately constant while their absolute scale may be altered almost without limit depending on distance or on the use of external instrumentation. Sharp diversity seen as part of a region on one scale becomes mere texture on another and eventually becomes entirely irrelevant except as it contributes to some average property. Though one could sequentially resolve all the details in a large scene, their relationship is rendered fuzzy by the fortunate failure of memory.

As in a physical aggregate, areas in the Kung Hsien painting can be distinguished not only by the nature of the individual brush strokes (=atoms) but also by the density and orientation of their arrangements. The trees in the forest are separable from each other by the orientation of their leaves, just as individual microcrystals in a pure polycrystalline material are distinguished by lattice orientation alone.

Each tree is unique; yet the recognition of qualities shared by trees gives style to them as well as speciation. But style is hierarchical; it resides at all levels, or rather between any inter-relatable levels. Considering the whole painting, its individuality and its style depend upon the distribution of and relationship between the styles of the smaller components. The painting itself can be said to be in the style of Kung Hsien because others by him are known with similar nuance of interplay between brushwork, pattern, content, and cultural overtones. On a coarser association it is recognizable as the work of a scholar-artist of the early Ch'ing period, then as Chinese, and possibly on a universal scale as earthbound and distinguishable from the styles that have developed in the works of beings on other planets. The painting itself as an undivided whole has become the unit for successive stages in the larger aggregate. The same sort of relationship appears in inverse form as one goes down in scale. Just as the areas were at first distinguished because of the patterns of brush strokes within them, so the brush strokes are recognized by the just-visible streaks and patches of light and dark on the paper, and these in turn derive from patterns of carbon particles (visible under a low-power microscope) that were deposited by the ink under the competing capillarities of the fibers in the brush and in the paper, while these in their turn arise in the molecular and electronic structure of colloids and polymers, and ultimately (or rather *pro tem*) in the patterns of interlock, mediated via photons, between electrons and atomic nuclei. One could move from here to any region among the atoms of any other solid without noting very much difference— until one returned up the scale of molecule, cell, crystal, and aggregate. The painting is just at the level at which the widening viewpoints of the artist and the narrowing viewpoints of the scientist merge. It is the scale of human experience, from which thought and imagination take off, and to which they must return.

Aspects of style are more than relationships of areas and volumes. The repetition of color or texture or of purely linear geometric detail can produce an aesthetic response in a viewer's mind. Note in figure 4 the unity that comes from the repetition of certain directions, curvatures, and angles of bend or junction which do not themselves interlock physically but are connected only in the viewer's mind. (The Japanese artist Sesshu [1420–1506] used such geometric resonance most strikingly in his famous *Four Seasons* scroll.) Moreover, the resonant pattern formed between the trunks of the larger trees in the foreground of figure 4 is different from that of the copse-like clumping of the smaller trees, but both serve to unify the whole on the larger scale of the painting. This extends still farther: though each tree is unique, recognition of a common dendritic quality relates them in the viewer's mind with all trees. Similarly, the individual houses in the painting are identified by the familiar closure of the pattern of lines, but the sight of them invokes the broader concepts of "housiness" in general.

Wang Wei, a painter of the late T'ang dynasty, delighted in emphasizing junction. For reasons evidently both philosophic and aesthetic, he played on the similarity between the branching connectivity of fissures in nearby rocks, the wrinkles and river valleys in distant mountains, and the branches of trees, the last, of course, the model for them all (as well as for modern computer programming!).

When the underlying structure is being examined the style of the whole is not visible, for this resides only in an external view of the whole. Bohr's famous principle of complementarity is, perhaps, nothing but a statement that things react on different levels. Our view is limited partly because we can neither see nor think of very much at one time, more fundamentally because the pattern of relationships is indeterminate until all the parts have been examined. George Kubler has said that style is like a rainbow. He is right because both style and bow depend on a constancy of relationships that exists on a scale below their own—the constant angle of refraction of light of different wavelengths in the case of the rainbow, and of a more complex mental refraction in the case of style. In both cases the phenomenon recedes as we approach the place where we thought we saw it, and is replaced by a previously hidden structure. Anthropologists can recognize a culture and analyze it, but are hard put to define it.

The eye finds beauty, even something analogous to style, in natural objects that have never felt the touch of a creative artist. In the last half century as artists have moved away from culturally determined iconography to abstract or supposedly nonrepresentational art they have unconsciously represented and have sometimes anticipated the natural forms that are revealed by advanced techniques in the laboratory. Actually, there are not many basic units of composition and as large things merge into smaller ones, and vice versa, both nature and the eye favor much the same principles of assembly. Landscapes, whether real or imaginary and whatever their scale or origin, have recognizable "style" based upon repetition, relationship, selection, and adjustment. Matter, whether living or dead, when left to itself adjusts its resonances on all scales, and the resulting structures seem to be based upon much the same relationships as those which give, or perhaps which actually constitute, aesthetic satisfaction in the mind of the human observer.

For five centuries, the advancing frontiers of science have been associated mainly with an increase in resolving power and the discovery of new levels of structure. In one direction we have found universes and galaxies, in the other we have found particles and subparticles, but having isolated them we still cannot understand their re-association except in rudimentarily simple ways which omit those qualitative features which give the richness that provoked human interest in the first place. Analysis followed by logical synthesis does *not* reconstruct reality. It leaves out local historical accident and balance, and all the cumulative complex consequences of individual history. A list of types of bricks used in the Hagia Sophia may help one to build an interesting brick wall, but it poorly suggests the great edifice from which they came.

Science in the past has been almost synonymous with distrust of the senses, but it seems to me that we are now ready to make better use of the other properties of the human brain besides its capacity for logical rigor. After logical methods that require an exact identification and control of boundary conditions have yielded what they can, we should seek a bridge to the more sensual study of whole systems. This will require an acceptance by scientists of macroscopic imprecision in the application of microscopically precise laws, and an appreciation of the individuality that arises historically in any complex system but has been excluded by current analytical or statistical approaches. We are not interested equally in all possible systems that could be formed from the units we know: those that *do* exist must take priority. Nevertheless, although atomistic details are insufficient for full understanding, they cannot be ignored. All the established "facts" must be considered before imaginative interpretation can be indulged in. Atoms and entropy may be dull to anyone except a physicist or chemist, but without them the whole rich world of thing and thought would not exist and there would be no one to enjoy it.

Entity and Entitation

Nothing is a thing by itself: it takes meaning, indeed existence, only as it interacts with something else. Once this is admitted, hierarchy becomes inevitable in all systems excepting only those that are completely ordered or completely disordered, for these have no levels between the units and the whole assembly. There are two unavoidable dissymmetries—moving upward, where the inner structure meets the surface of whatever we are talking about, and, moving downward, where it meets the surface of a part.

The very fact that something is recognized as an entity means that it has associated with it a distinguishable response to external probing; in physical terms some measurable property, in more general situations some recognizable quality. The hierarchical alternation between an externally observable quality, property, or trait and an internal structure which gives rise to it occurs at all levels and applies to all things. The interface, being both separation and junction, is always Janus-faced—that which is characterized as an entity is in some way more closely connected within than without. To define an entity, to separate it from the rest of the world, the interface closes upon itself, and hence in sum, but not necessarily everywhere, is concave inwards; connections converge and tighten inside, at the same time that they diverge and loosen outside it. Gertrude Stein once remarked that identity, based upon interaction, was much more interesting than entity, which marks completeness. Moreover, as Alice's whiting said to the snail, "The further off from England the nearer is to France."

Though all structures depend in the main on the repetition of relationships, there is always some hierarchical level in any natural, social, or aesthetic structure at which it can withstand the replacement of some of its parts by others. At what level are anomalies or misfits rejected, tolerated or welcomed? In some structures the simple hierarchical concept of the whole selects and places the parts leaving little possibility of interchange without destroying the whole. In others, the parts rigidly fix the pattern but not the extent of the whole, as, for example, the lattices of hard polygons in a Moorish mosaic in which no random change is possible, except at the edges, without disruption—though color remains a variable to be played with by adjustiment and substitution within modules maintained at a higher level, as in pieces on the squares of a chess board, or atoms in alloy crystals as we discuss below. It is also common for misfits to compensate each other and to satisfy the balance of connectivities within larger modules. Thus, pentagons, which when regular cannot fill space do so when grouped in pairs to form quadrilaterals or in fours to compose hexagons (figure 5). Any extended array, however irregular or lacking in symmetry it may be, can always be subdivided into space-filling groups within boundaries which connect in quadrilateral or hexagonal array.* The scale at which this can happen is an index of the hierarchical nature of the system. The balance or imbalance of internal and external connections determines both identity and the possibility of extensive aggregation.

When an extended array is viewed with decreasing resolution on larger and larger scale, the system becomes simplified by the merging of compatible connections, turning polygons into their duals (points with valence equal to the original number of sides) and groups of polygons into simpler polygons and eventually into points with the same residual valence. Thus up the hierarchical scale forever as local differences merge into global balance.

Such requirements are very simple applications of Euler's law relating the numbers of 0-, 1-, 2-, and 3-dimensional features in a simply-connected group of polyhedra. Being topological this does not relate in any way to shape, only to contacts and separations, but when they are combined with simple physical interactions interesting geometric effects appear. Thus in a random soap froth in two dimensions (figure 6), the necessity of compensating for cells having more than six sides with those having less than six is joined to the surface-tension requirement for minimum boundary length and junctions at 120°. The cell walls therefore have to be curved, and they can be rendered stable only by pressure differences between adjacent bubbles. This, in turn, causes slow diffusion of gas from small bubbles into their larger neighbors and introduces perpetual instability.

The limitations represented by Euler's law make it impossible to have extended simply-connected aggregates of polygons in two dimensions with greater valence (number of neighbors) than six, or convex cells in three dimensions with more than fourteen: the number of neighbors needed to satisfy higher valencies cannot be assembled repetitively into a connected aggregate. However, a one-dimensional branching tree with no return loops can be of any extent and degree of complication. In three dimensions, such a tree can be externally superimposed on a two- or three-dimensional array to connect together any number of cells whatever. This, of course, is the basis of organism, for it provides a means of communication that transcends neighborhood: aggregation and communication are no longer limited to interactions only with, or via, nearest neighbors. The human mind similarly establishes relationships between parts of systems both external and external to itself which do not necessarily have any physical inter-

*See C.S. Smith, "Structural hierarchy in inorganic systems," in L.L. Whyte et al., eds., *Hierarchical Structures* (New York: American Elsevier, 1969).

connection. Thought is supposedly a kind of trial of patterns of interconnection of neurons as they already exist within the brain and as they are modified by selective interaction of the structure of the whole body with the outer world via the senses of sight, sound, touch, taste, and smell.

The distinction between an inorganic aggregate and a viable organization or organism lies mainly in the fact that the latter contains an essentially one-dimensional tree-like system maintaining communication to and from a center of some complexity. The communication system, being one-dimensional, cannot maintain itself unless it is immersed in a three-dimensional aggregate, but it often provides the means whereby the nature of the 3-D parts can be changed and their capacity for interaction with their neighbors thereby altered. Communication may be directed or broadcast, but there must be eventual feedback if the system is to be maintained.

Without both tension and compression and the balance between them nothing could exist, for it would either expand to infinity or shrink to nothing. Though tension can be one-dimensional, compression, as every engineer knows, has to be three-dimensional if pressure is to be contained or buckling avoided. The number of restraints that are needed to prevent buckling is, indeed, a mark of the dimensionality of the system. The part in compression can be internal within a skin stressed in tension, as in a soap bubble, or it can be external, as a tetrahedral framework of rods supporting four stretched wires meeting at a central point or in the more complicated "tensegrity" structures of Buckminster Fuller and his followers. Even the brain needs a three-dimensional structure underlying the sometimes two-dimensional patterns of thought. A structural entity is one in which tension and compression are internally self balanced. (Two-dimensional systems can achieve stability in a three-dimensional world by the balance between centripetal and centrifugal forces as in planetary systems, the bolas, or the spinning electron—though note that the electron cannot interact with anything else without the intervention of a photon having an internal quantized structure that is at least three dimensional.)

Few entities are completely isolated without some residual connections to others within a larger aggregate within which either tension or compression or both form cell walls, simultaneously dividing and maintaining continuity. It is the stability of an overarching 3-D environment that allows special freedoms and local structures, for example, the holding of feebly-bonded atoms within the chelated structure of some molecules, the tolerance of a crystal lattice for vacancies, dislocations and substitutional atoms, and, most important of all, the balanced interlocking of separable

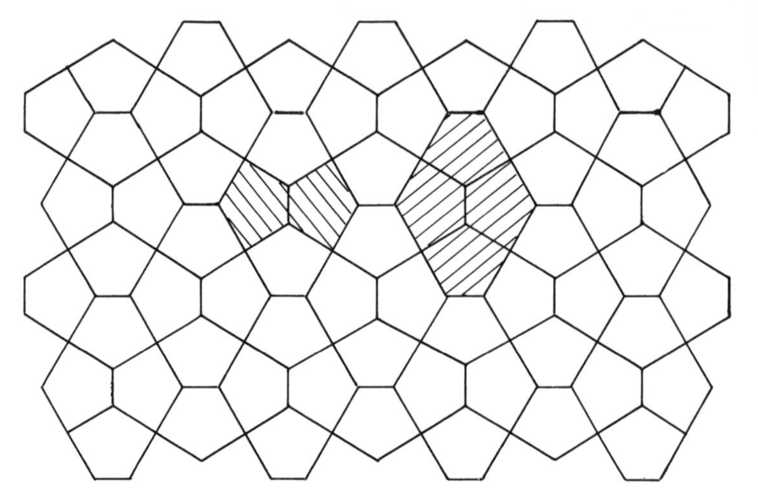

Figure 5. Space-filling arrays of pentagons in hexagonal and quadrilateral tesselations meeting, respectively, at trivalent or quadrivalent vertices.

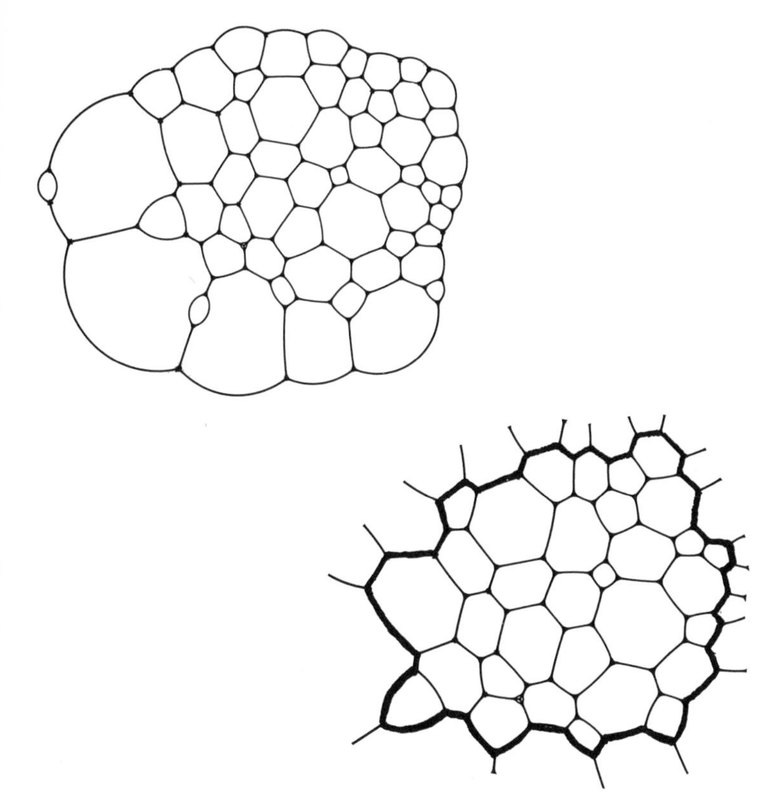

Figure 6. Soap bubbles in two-dimensional array. Since all corners are trivalent, the average bubble in a large aggregate must approximate six sides, and in an array of any size the departure from six is detectable by inspection of the boundary alone.

In an isolated set such as (a), with all vertices three-fold and with no dangling edges, then, if P_n is the number of polygons with n sides and E_b the number of unshared edges constituting the boundary,

$$\Sigma\,(6 - n)P_n - E_b = 6.$$

Note that this is not the average but the sum of departures from hexagonality. Similarly, if a group of such polygons is excised from a larger array, as in (b), then

$$\Sigma\,(6 - n)P_n + E_o - E_i = 6,$$

where E_o and E_i are, respectively, the number of edges extending outward (severed) and inward from the boundary. Similar relations between internal and external connections apply in all systems, but they are only simple in two dimensions and when all vertices are restricted to the same valence.

molecules with homopolar bonds, hydrogen bonds, and van der Waals bonds that lies at the basis of biological replication. In solids, the bonds between atoms may be as sharply directed as the covalent bonds between tetrahedral carbon atoms in diamond, as specifically neighborly as in an ionic crystal, or as regionally compensating as in a metal.

Substructure

When more than one type of "atom" is involved, many different hierarchies become possible within the same overall lattice of principal connections. A part can be an unchangeable "atom" for some purposes and a variable organism for others.

This kind of structure is well illustrated by the simplest binary alloy incorporating equal numbers of two kinds of atoms, shown schematically in figure 7. Preference for dissimilar neighbors leads ideally to perfect order, figure 7a, and this is observed in crystals of many ionic compounds such as sodium chloride where the energy penalty for wrong neighbors is high. With less dominant neighborly interaction there may be no order at high temperatures—the random solid solution of figure 7b—but on cooling order begins slowly and gradually increases. Here it is easy for local regions of order to be out of step, producing a zone of misfit (figure 7c), a domain boundary somewhat analogous to the orientation misfit of figures 1 and 8, but far less costly energetically. The same occurs with ferromagnetic interaction, where interlock of local magnetic polarity due to electron spin orientation defines regions that (except in the very smallest crystals) compensate each other in the whole. Changes into structures of this type, occurring within a dominant framework that is not significantly altered, are the second order transformations of the thermodynamicist. They are called cooperative phenomena, but they could just as well be called anticooperative, for local cooperation produces larger scale opposition: it depends on what scale is observed. Diffuse interfaces are common.

An adequate description of a system must include not only local neighbors and the average of the whole, but also must convey a sense or measure of the hierarchical interlock between statistics and scale. Imperfections in local ideal arrangements inevitably arise as the structure on any one level is extended, and those imperfections point to and permit the existence of a higher level of structure. This continues until eventually, on a sufficiently large scale, *all* segregation, order, disorder, and diversity become invisible. My colleague John Cahn has called attention to the fact that even such a simple property as chemical composition has hierarchical aspects. What is the composition of a solution? For example, consider a crystal of beta brass, an alloy of copper and zinc containing about equal proportions of atoms of the two elements arranged in structures rather like those in figures 7a–7d depending upon temperature. Obviously if one takes for analysis a sample the size of a single atom the composition will be 100 percent of either copper or zinc, while a smaller sample might fail to contain any nucleus whatever. With a sample volume large enough to include two atoms, both could be of either element, or a one-to-one association of copper and zinc could come up, indistinguishable from the nineteenth-century ideal molecule of a compound CuZn, whatever degree of order or disorder might prevail on a larger scale. This hierarchical-statistical quality is quite general and applies to the component parts of anything whatever.

It should be noted that if there is more than one component in a structure each of the smaller ones can, independently of the others, take its place in its own array with the characteristics of a solid, liquid, or gas. In a gas all things are miscible, but gas-like distribution can also occur of some atoms in a crystalline solid. An ordered framework of one component may or may not be accompanied by order in the others, although the smaller components cannot have more order than the larger ones. In felspar, for example, within a persisting lattice-network formed by the negative ions of oxygen, the major positive ions of silicon and aluminum and the smaller positive ions of potassium, sodium, and calcium can be substituted for each other either randomly or each one in separately ordered positions to an extent limited only by the thermal history, the size of the atom, and the requirement of electrical neutrality. Of course, there is strong interaction between atoms which, locally, can affect their neighbors just as an external pressure would. Some components are more easily substituted than others, and some subgroups form lattice complexes or radicals that have to be moved intact. Analogies to all this are easy to find in the structures of art and social institutions.

Figures 7, 8, and 9 illustrate an important structural feature that can only exist within higher-order, namely vacant, lattice sites. The atoms adjacent to these vacancies are restrained by the normal interlock of their environment, but an atom and a hole can interchange positions easily with no immediate long-range opposition or effect. Lattice vacancies are as essential to the physical world as the crystals within which they form. They are the basis of diffusion in solids, for they allow movement of units by local interchange under thermal agitation without destruction of the overall pattern: the system can move toward either segregation or randomness depending on the statistical bias in the interchange of neighbors as vacancies move about. It should be noted that such diffusion affects only the units of the structure, not the superstruc-

Figure 7. Lattice composed of two different kinds of atoms distributed randomly (a); in ordered array with each atom surrounded by the opposite kind (b); in two regions of order with an out-of-step boundary (c); with a diffuse gradient in composition (d); and with complete segregation (e). The basis of a boundary outlining structure on a higher hierarchical level is present in all but (a) and (b).

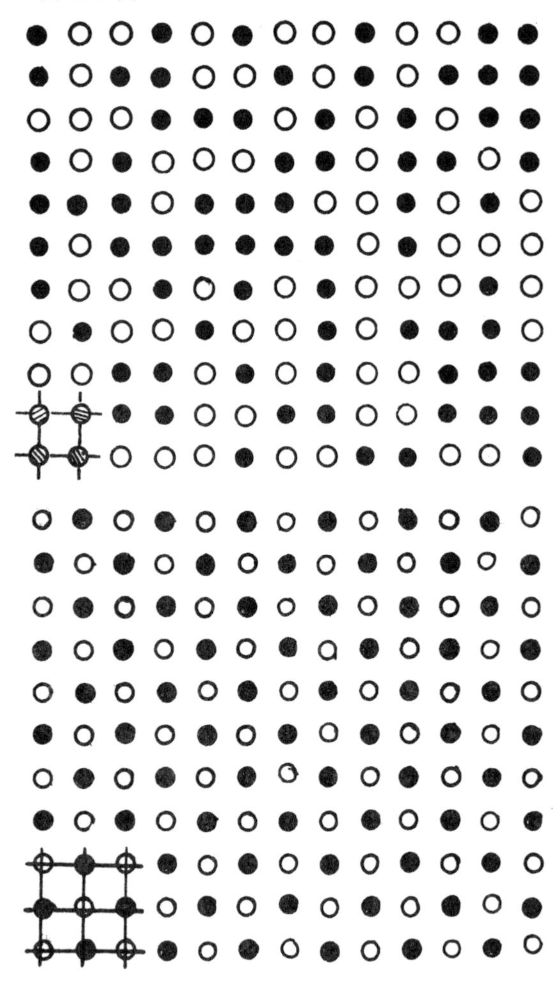

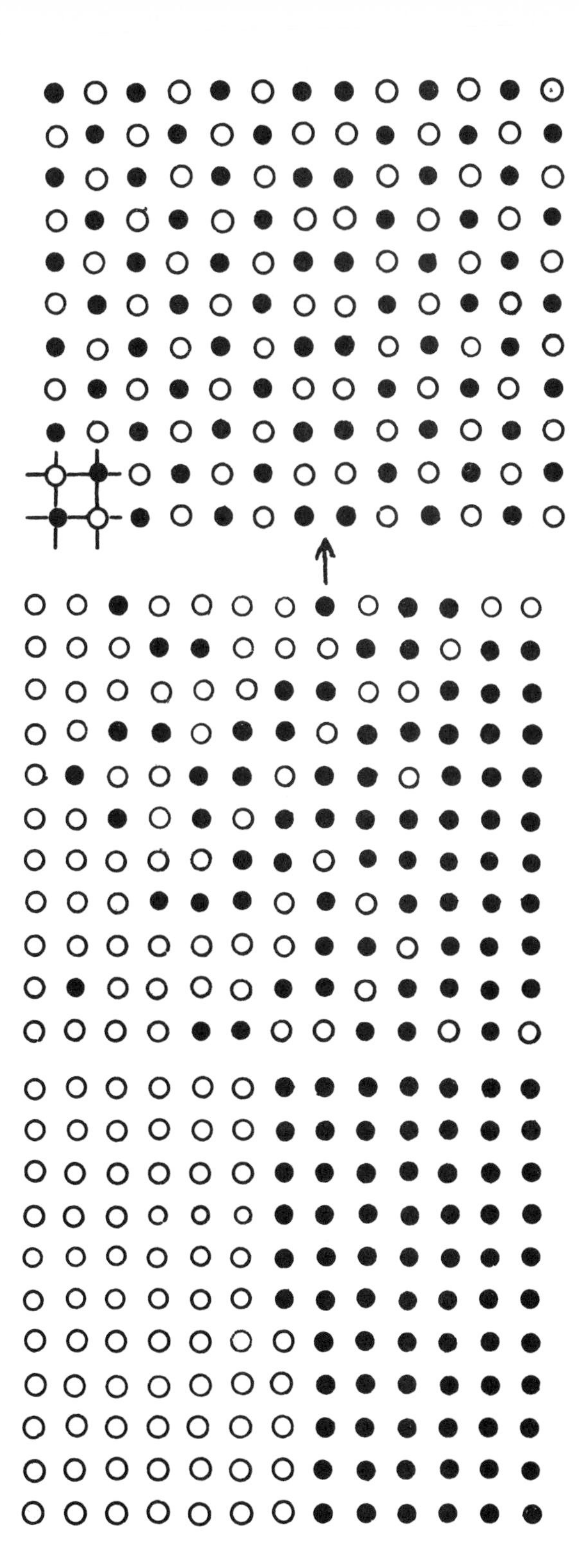

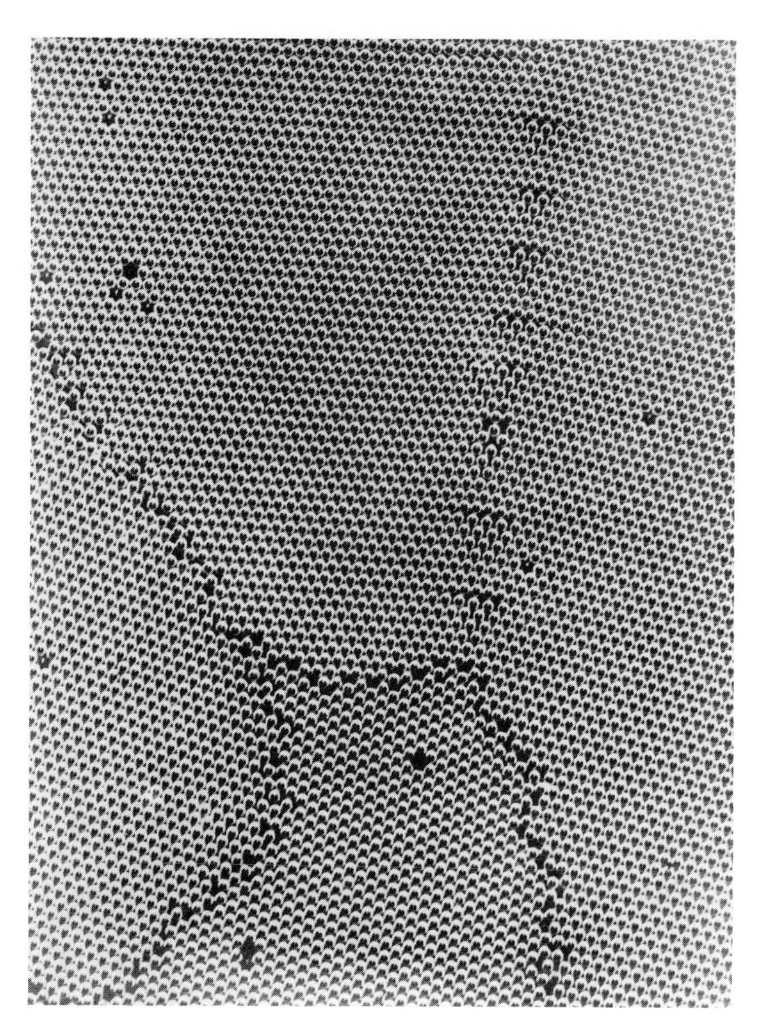

Figure 8. Array of tiny uniform soap bubbles (ca. 0.5 mm diam.) serving as a model of a region in a simple crystalline solid, with vacancies, substitutions, dislocations, and lines of misfit (boundaries) between areas differing in orientation. From Metal Progress; © American Society for Metals, 1950.

Figure 9. Diagram of a "square" lattice containing four places where atoms are missing without disturbing the larger arrangement, and a small region of misfit where the atoms are arranged in triangular array.

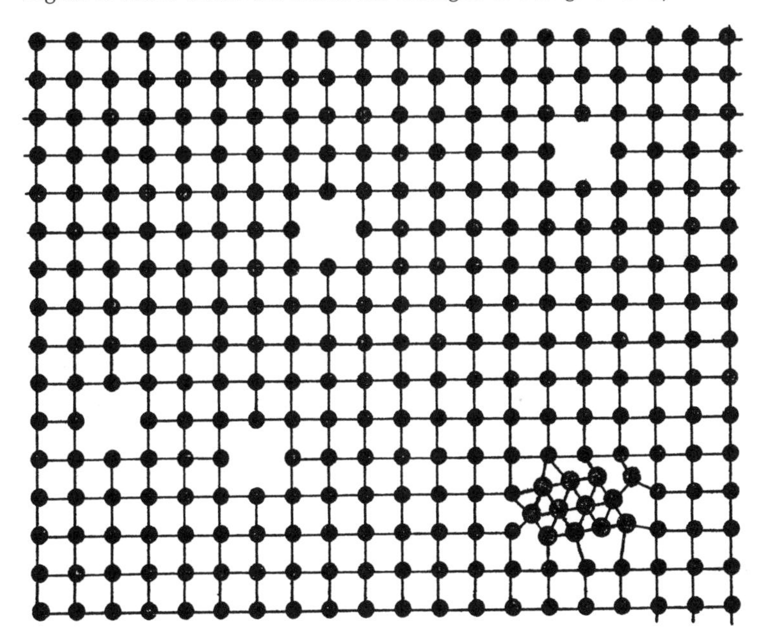

ture itself. However, although diffusion may occur by easy stages within an existing framework, the resulting local changes in the nature and concentration of units may set up conditions that will feed a revolutionary reconstructive change once a nucleus appears.

As in matter, so in the structure of ideas. "The thing that had happened in the heart" is a necessary preliminary to political revolution. In the history of science, one sees similar patterns of interlock followed by gradual change preceding the structurally reconstructive changes that constitute the paradigm shifts of Thomas Kuhn. Dalton's simple model of the molecule made nonstoichiometric compounds invisible to chemists for half a century, while physicists were unable to deal with the properties of real materials until after the discovery of disorder, which in turn was only possible after a period in which first atomic and then crystalline order had been found and overemphasized. In the last half-century departures from older norms of crystallographic symmetry and ideal electronic energy levels have been recognized, causing a complete revolution in the concept of materials and an explosive growth in the utility of solid state physics.

Mechanisms of Change

This discussion is intended to help those familiar with the material sciences to see a little deeper into the quality of art, and those who enjoy art to understand a little more about the nature of substances. The recognition of the style in a work of art or a school was likened to the identification of a chemical phase, for both require the recognition of a more or less extended area characterized by certain qualities that inhere in the structural arrangements of parts on a lower hierarchical level. The differences lie mainly in the number of distinguishable components and in the character of the external observation.

The similarity between phase and style extends beyond mere existence—it can be seen even better in the mechanism of change that occurs to new forms when modifications of conditions have rendered an existing structure no longer the most appropriate one. Change of both phase and style occur by the regrouping of parts (which are usually far more stable than the whole) into a new pattern of interlock; a pattern which is fortuitous and highly localized when it first appears, but which provides the opportunity for self-discovery of the fact that a new grouping is preferable to the old, and is followed by an increase in the amount of the new at the expense of the old until the latter is consumed. Without a tiny sample of the new structure to serve suggestively as a nucleus, the old will remain forever happily metastable, for it has no means of knowing that change is desirable. Thus, crystals of ice melt not because they are suddenly less stable at +0.01 °C than they are at −0.01°C, but because the whole system benefits by the motion of the interface between ice and water. The interior of the ice does not distintegrate or change in any way: the volume of water simply increases at the expense of ice as long as the necessary heat is supplied from outside.

A system that is in equilibrium under given conditions has reached balance between the requirements of different levels, and a subsequent change in any level will eventually influence others, but at different rates. Because parts adjust faster than wholes, change moves structurally upward. In an equilibrated structure change will not begin unless there has been an externally induced change in conditions—something equivalent to the relative numbers of different species of units (the composition), or their intimacy (density, reciprocal of pressure), or the frequency or form of communication (temperature, number of quantum states or larger groupings of information). A change will affect at least four levels, that of the changed entity itself, that of its environment, that of its internal structure, and that of the medium of communication. A change in any of these may make a stable structure become metastable, ripe for change as soon as a nucleus appears to suggest a better arrangement. In an atomic aggregate different atoms do not, of course, form internally (except in the rare case of nuclear transmutation) but something akin to this occurs easily in more complex systems: the parts themselves being complex can individually and locally undergo internal transformation. If enough similar ones form as a result of suggestive communication, the replicated new units (which do not need to be identical, only reinforcingly communicative) will find each other and jointly respond by reformation when an appropriate nucleus of a better superstructure appears. This is the way ideas spread and topple governments, or, in science, Kuhnian paradigms succeed each other.

Not all changes, of course, involve the complete reorganization of the environment. There are many cases of small nonrevolutionary change that occur in a lower level of a pre-existing structure without straining the local connections to the point of disruption. Inventions are improved by bright ideas in every mechanic's shop, and every member of a school of artists helps to round out the master's style in popularizing it. Such is the very essence of ecological gradualism and has strong implications for biological and social development, because change can begin by gradual substitution and without discontinuity. This may stabilize the larger structure, or it may gradually set up conditions that make disruptive change probable. An example of the latter is the easy acceptance of technological change on an individual basis—the automobile, telephone, or dishwashing machine, for example—eventually interacting to make large social change necessary. This is analogous to the slow substitution of, say, chromium atoms for iron in a face-centered cubic lattice at 1000°C by diffusion, eventually making a body-centered cubic structure preferable and waiting only for a nucleus to produce it.

Any existing structure possesses structural inertia. Because of the reinforcing interactions of its parts a structure resists change even after it has become thermodynamically, philosophically, or socially metastable. It came into existence because at some time in the past the parts had found stable patterns of overall interlock, and the interaction of any one pair of neighbors cannot now be altered without altering many adjacent ones. Anything but an isolated structureless particle or the nonresonant chaos of a gas has some history locked into it, and the more complex a thing is the more its present features depend upon the retention of unique configurations that resulted from the resolution of some historic conflicts during growth. Minerva's full-panoplied origin is a material impossibility, but the myth itself has permanence because of the long conceptual history behind it.

Simple structures allow few opportunities for the incorporation of alternative arrangements as they grow, and their history is without much interest. Conversely, the local groupings embodied in a complex structure both record past events and provide a unique framework on or within which future changes must occur. Though overarching order may erase some details, local differences that mesh into the larger structure will be preserved and built into the future.

Some time ago—appropriately on Memorial Day, 1970—I coined the word "funeous" for this aspect of structure, after the unfortunate character in Jorge Luis Borges' story, "Funes, the Memorious" who remembered everything. Different structures have widely differing degrees of "funicity." Physics is largely afuneous. In general, both the analytical approach of the physical scientist and the averaging methods of the statistician achieve their exactness by the elimination of funeous aspects of the world. However, it is both impossible and unnecessary to study all structures that might have come about in the developing universe, and in the future science will inevitably pay more attention to those complex funeous structures that actually do exist. History, which produced the record, and physics, which analyses it, must work hand in hand. Biological organisms and human cultures arising therefrom have a high density of funeous detail, indeed their very nature depends on the transfer of blocks of historically acquired pattern, not statistical exploration at each point of change in the development of each individual. The heredity-versus-environment argument can be applied to all things, for the past has given structures at various levels that cooperate to respond to present opportunities.

Man is at the scale at which structures built of chemical atoms have achieved something like the maximum significant degree of funicity, but his discovery of new means of preserving, replicating, and communicating blocks of thought patterns makes new levels possible. By using only the properties of pattern in electronic communication, thought can extend immeasurably far beyond the limits imposed by the aggregation of matter. (It does, however, need a material substrate on which the patterns can be formed.)

In the generation of new structures, that is, in change, the ordering of the parts has to pass through a period of explorative contacts and the gradual adjustment and locking in of the slow responses with the fast ones. Funicity, that is, historical diversity, is rooted in quantized interaction of structural units at various scales of interlock though ultimately on the energy quantum. Quantum mechanics is the basis of all stability. Too much emphasis is commonly placed on the principle of uncertainty in quantum physics, which does not affect the result of an interaction, but only whether or not a particular interaction occurs within a given interval of time. Perhaps time itself is nothing but sequence in the hierarchy of structural inertia. The various quantum levels within the atom represent the patterns of resonant interlock between different ratios and densities of nucleons, electrons, and quanta, exactly as, on the next level of hierarchy, atoms in different ratios interlock to form a series of chemical compounds in Daltonian molecules. Larger cooperative structures such as crystal aggregates take longer to form because the response time to find if resonance has occurred is not based on the speed of light, but requires diffusion to relieve strain in cooperative configurations.

Once a structure has found itself, the integrity of the whole stabilizes the parts and resists their rearrangement. There is a close interplay between the numerical and the morphic aspects of any system. Structural inertia is at least as important as mass inertia, indeed it is the basis of inertial mass in everything larger than an electron. Macroscopic behavior predicated upon Newton's third law (action = reaction) occurs only when the interacting bodies are not stressed beyond the point where their internal structure ceases to respond elastically. (This is why physicists in the eighteenth and nineteenth centuries studied only the mechanics of elastic bodies and left plasticity to the potter, the pastry cook, and the metallurgist.) Reversibility applies only to binary interaction or to an elastic assembly. Consider the difference between the action of a hard steel punch and the reaction of a piece of soft silverware being chased. They are anything but equal and opposite except on the subatomic scale. The classification of actions as either elastic or inelastic is quite fundamental. All structures, and parts of structures, have a kind of elastic limit in their response to external conditions beyond which relaxation is to a pattern of connections that differs from the initial one. Without this there would be no historical individuality in complex structures.

Any change whatsoever involves a catastrophic change of neighbors at some level, while both above and below this level connections, though they may be strained, remain topologically unchanged. Atoms persist through changes of state and combination, while the local fury of a storm leaves the global atmospheric balance unchanged. Statesman can talk of destroying governments but not people, and the death of one man does not immediately change either society or atoms.

Nothing is unchangeable unless its environment is also unchanged. The only ultimate truth is, "It all depends." However, within a complex system not in equilibrium local differences in the rate of loss of memory (that is, the erasing of structural features formed under conditions that no longer exist) will continue to cause internal change. A crystalline bar can be bent by an external force (fig. 10a, b), but if it is elastic it recovers as soon as the force is removed: The strain caused a change only in the patterns of quantum/electron resonance forming the bonds, not in the topology of the connections between atoms and their neighbors. At a sufficiently high stress irreversible plastic deformation becomes possible, but this of necessity involves a topological change; the production and movement of some local imperfection in the symmetry of the orginal crystal lattice. The imperfections (fig. 10c) known to the solid state physicist as lattice dislocations, have as much geometric reality as the groupings in the perfect lattice, but they can exist only in an extended environment of such perfection. They are hard to form but, once formed, are easy to move. In the simple two-dimensional case of figure 10 every "atom" has four neighbors except at the dislocations in figure 10c where a single atom has three neighbors and all connections outside its immediate neighborhood are unchanged except for a regional change in direction. Thus, a few places of concentrated misfit have absorbed nearly all of the applied strain and the rest of the structure has been allowed to relax. The energy of the entire system has been decreased by replacing the sum of many small strains extending throughout a large region with the extreme strain of a local catastrophe. The bent bar no longer unbends on the release of the external force. The plasticity of metals, like the possibility of social change, depends entirely upon the generation and movement of imperfections within a predominantly ordered environment.

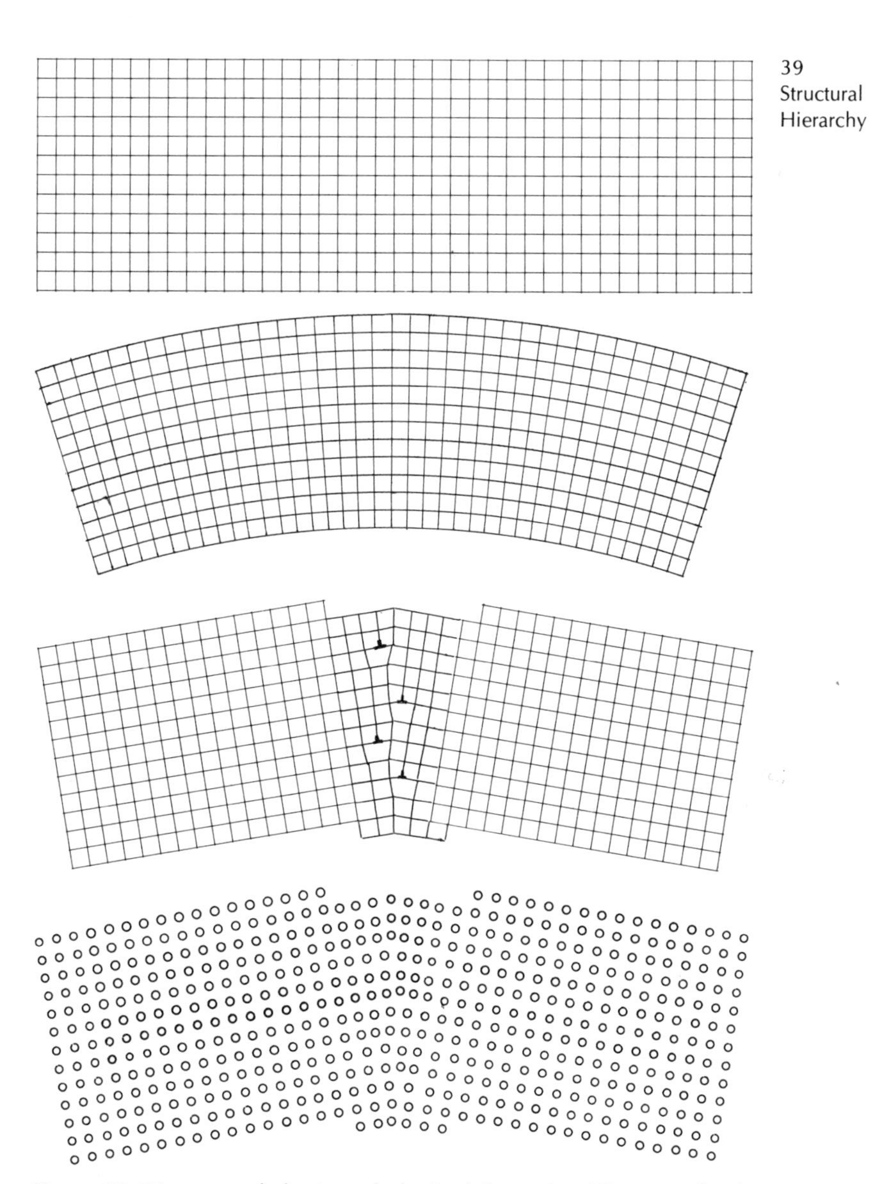

Figure 10. Diagrams of elastic and plastic deformation. The unstrained crystal (a) can be bent by an external force (b), but will return on removal of constraint. However, if local imperfections ("dislocations") are introduced, the crystal will remain bent while leaving most of the lattice relaxed (c).

Both the nature of the parts and the nature of the communication between them are necessarily involved in the interactions that stabilize structures, and changes in either may render the superstructure metastable. Internal structural change may also be rendered desirable by an externally produced change in volume or in temperature, or if the atoms themselves are changed as a result of nuclear transmutation of diffusion from outside. However, even if a new array would have lower energy it cannot form instantaneously, for the new structure must be nucleated in an environment that is opposed to it, and it must find a mechanism for growth.

A small group of units which, if isolated, could easily move into a new configuration cannot do so when it is embedded in a larger structure, for the initial coherence with the environment produces stresses at the interface that are equal and opposite to the displacive ones until some atoms switch their connections. The initial elasticity hinders the discovery of the possible advantages of any new structure. However, the advantages of the new phase increase in proportion to its volume, and there is some critical size beyond which the energy required for the interface is more than compensated by the volumetric gain. So with things in general, for a "thing" exists only in interaction with its environment, and it includes the interface through which it merges with or adjusts to its neighbors.

In the arts, once stylistic norms are established, slight departures are welcomed as improvements, but really original deviants are extinguished or at least ignored. At some point, either because of boredom or because of long-range changes of mental attitude (perhaps as a result of an intrusion of forms from another culture), social tolerance for at least some kinds of novelty will increase; the innovations will become numerous enough to reinforce each other and so overcome the conservative pressures of the status quo to form an effective nucleus. As more and more patrons accept the new form, more and more artists will produce works in it and it becomes a recognizable style with its own opposition to further change.

Look at figure 9, bearing in mind that it is two-dimensional and overly simplistic. It depicts an area of a two-dimensional crystal based upon a square unit cell, each "atom" being joined to four others. It contains, however, a local region near the lower right corner in a six-connected triangular arrangement. Now, even if this should be a preferable structure and even if there should be no change in volume, its formation is opposed by the whole existing array of established connections between neighboring atoms, and both atom positions and a whole hierarchy of internal and external bond resonances must be disruptively changed to allow the formation of the six bonds needed for each atom in place of the original four. Once the particle has become large enough so that the energy associated with the disruption at the interface is less than the gain within the volume, the system will move in whatever direction favors the structure with lower free energy. This kind of interface is of necessity disordered, and atoms can pass into and out of it with little more restraint than if it were liquid. Its structure permits and its translation produces the desirable change of arrangement that could not occur within the uniform lattice of the adjacent material.

When a change of temperature, pressure, composition, or social climate has made a system ripe for change, nuclei of a more stable form will not appear everywhere,but only in a few places within the old structure that are for some local reason deformed or strained, or at an interface with an intrusive structure that serves to catalyze the change with a suggestion of a possible new order. In a physical system of crystals, new forms are most likely to appear in the zone of misfit where one phase or crystal impinges upon another. (Note the many such regions in figure 8.) Socially, it is where individual freedom is greatest, usually at places where classes or cultures clash ; aesthetically, it is where ideas least conform to the established values or where an existing style is impressed on a new material or a different technique. However, these same structural suggestions can exist locally without nucleating massive change as long as the dominant structure best matches the prevailing "thermodynamic" conditions, which are overall, not local.

One cannot overemphasis the fact that everything—meaning and value as well as appropriateness of individual human conduct or the energy state of an atom—depends upon the interaction of the thing itself and its environment. The drive for both stability and change is the minimization of free energy in a physical system and, in a social system, something like unhappiness or dissatisfaction; both summed, not averaged or individually measured. The mechanism of change is by transfer across the interface, each atom making its own choice, and quickly being brought back if not supported by compatible choices of its neighbors. Not infrequently, such short-range improvement delays an overall readjustment that would be better—the metastable phase or the tragedy of the commons. Is loyalty to a country or belief in one system of moral values admirable when it denies equal value to others?

In a large system local changes may be nucleated in many locations, and patches of the new, more stable, structure will continue to spread until eventually they impinge and interfere with each other. Once the difficult stage of nucleation has been passed, patches of the new form will grow by accretion at a rate depending only on how quickly the parts at the interface can rearrange themselves. This is the central rising slope of the now-familiar "S"-curve of growth, figure 11. The slowing and stopping of growth comes from the depletion of material to be changed into the new form or from the concentration of rejected parts that cannot be removed by diffusion or otherwise. So with salt in water that is freezing or, less permanently, heat in almost all reactions and conservative ideas in a radical society.

Change, then, begins slowly, uncertainly, and in places that are highly dependent upon local circumstances because the nuclei necessarily are misfits in the existing structure or orthodoxy. The nuclei are unpredictable (except perhaps by the Witch of Endor of whom Banquo demanded "If you can look into the seeds of time, and say which grains will grow and which will not, Speak then to me") because no system can by itself know ahead of time what, if any, new structure can supplant it. Nuclei do form, however, in those regions of the old structure that are least contented. A phase change is analogous to a political revolution; not the destruction of all individuals but the rearrangement of most of them into a new pattern of interaction. A revolution, driven by the injustices of the old regime, needs its formative nucleus, and its growth (which occurs via an interface of high disorder) is slowed by the need to accommodate or eliminate dissenters. Eventually, however, the structure that began as the highly creative work of a successful innovator becomes an ideology and as it spreads it becomes indoctrination, not creation.

At the top end of the "S"-curve of growth where the new structure has in the main replaced the old, there follows a period of ripening, of adjustment in response to smaller energy changes. Personal preferences begin to modify a dominant but overly simplistic ideology; necessity at last becomes the mother of invention and mechanics can improve mechanisms; a school of lesser artists refine the creator's insights and methods; local regions of stable minor phases are formed by the segregation of "impurity" atoms and the breaking away of clusters of them from the dominant older matrix; and large particles of the stable phases absorb smaller ones to decrease interfacial area. The system slowly moves toward a stable maturity that will be upset once again only if something from outside changes the nature or density of the component parts or their means of communication. While at the beginning a small nuclear region expands to create large change in an unstable structure, at the end the stability of the new larger structure dominates the smaller details. There is individual injustice in both cases. The conflict of large regions of order produces local conditions of disorder that can neither extend nor vanish for

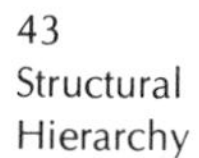

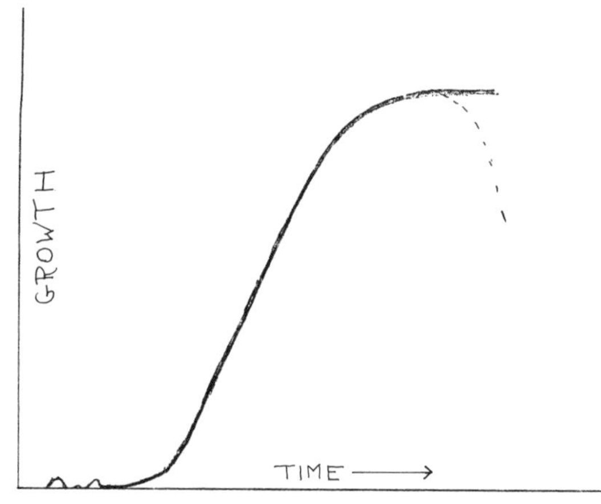

Figure 11. The "S"-curve representing the growth and maturity of anything whatsoever. Nucleation is rare, local, and difficult, but once growth has started it proceeds for a time without hindrance until the point is reached where depletion and conflict cause growth to slow and stop. Thereafter further change is by slow rearrangement of internal anomalies introduced by accidents of growth.

Figure 12. Ceremonial axe of nephrite, cut and polished by the use of abrasives. Chinese, Shang (?) Dynasty. Courtesy of the Fogg Art Museum, Harvard University, No. 1943.50.130.

their dissymmetry is in local equilibrium with the overarching stability of the larger environment: They shrink but cannot disappear. This contrasts with the beginning, where it was the environment that was unstable and the interface moved outward from the misfit. What was once an innovative individual thing has become just one part of a larger structure which now claims the individuality.

The total effect of these overlapping three stages in the growth forms the "S" curve. Both beginning and end are gradual and both depend on virtually unpredictable local conditions. Though it is easy to follow the growth of a new style or a new phase, for it is simply a quantitative change, it is virtually impossible to observe its beginning or even retrospectively to trace back to it. The creation of a form cannot be observed until it exists, and, in practice, it is not usually noticed until after it has existed for some time. The only certainty is that the beginning of a new structure must be a misfit with the old. New political ideas, new aesthetic forms or new scientific theories inevitably seem crazy in the framework within which they appear. The converse, however, is not true—it must not be assumed that all misfits have the germ of the future in them! In physical systems most fluctuations are transient and in human societies the overwhelming majority of crazy ideas remain crazy forever, in any environment.

Any historian who has tried to trace the origin of an important concept or invention knows how difficult it is to do so, for even the people most concerned are unaware of the full consequences of what they are doing. Even in their own time, the records of the beginning stages of any art, technology, language, or other component of culture are few in number and hard to detect, and even when a change has proved itself by incorporation in the social structure its origins are cherished only in retrospective, perhaps overly mature, societies. On the other hand, the period of growth leaves many records and is driven by the very type of social consensus that had earlier opposed it. Change is in one direction only, and people at the interface must chose between old and new, being in contact with both. The period of maturity again has many locally diverse features, unpredictable because they result from the interaction of previously unconnected events, from the accumulation of rejected or unexpected factors, and from the reconcilation of larger conflicting domains. Even the most careful design or advanced systems analysis can rarely include all important consequences of a new activity. "Technology Assessment" cannot hope to anticipate all the side effects of a new technology, through it may be able to detect them early, to avoid catastrophe if they are harmful, and to obtain greater benefits otherwise. The aesthetic imagination of the artist has little role to play in the middle of the "S" curve but is essential at both ends, as it suggests new structures that are capable of expanding or new arrangements that best resolve local tensions arising in the conflict of larger stable structures. If a means of replication exists the latter may in their turn become units in a structure of society on a still higher hierarchical level. The fact that the social consequences of art are not easily seen at first makes it easy to spread on a personal level.

This model of nucleation, growth, and mature adjustment, simplistic though it is, may illuminate some anthropological problems. It is only after mass acceptance and interlock with other aspects of behavior that a trait can be regarded as characterizing a culture.

As we have seen, the interactive quality of structural inertia makes the formation of a new structure extremely difficult. If it should occur once, the chance of a similar nucleus occurring independently at another place remains not a bit greater than it was before. But a new structure can, of course, spread into other parts of the old by the example of direct contact, or by word and diagram carry just the idea to catalyze comparable change in other receptive areas. The stimulus has no power to force change but merely offers an example of a pattern the response to which depends entirely upon local conditions. The difficulties of "modernization" programs and the transfer of technology to "underdeveloped" countries show how different levels of a structure may be transferred at different rates, producing curious hierarchical disregistry. In fact, except by conquest, only those parts of a culture or civilization that are capable of being reintegrated into a different larger pattern are transferable. Cultures are the hierarchical result of collective selection—a phrase I owe to Avraham Wachman.

Personally, I believe that human communication in some form is behind nearly all cases of seemingly independent simultaneous invention, the multiple appearance of ideas or concepts that are outside the pre-existing framework—things like the first smelting of ores, for example, and perhaps first ideas of kingship and of God. That the route of communication does not always remain visible is a consequence of the selective unreadiness of the intermediate terrain. An ice crystal moving over a moist windowpane will leave no structural record except in those places below freezing temperature, and travelers' tales are usually disbelieved.

The changes involved in the refinement of a mature system are different. The adjustment of the larger structure tends to force a pattern upon the localities of conflict, and rather similar structural resolutions may occur in many different places. These are the "reverse salients" of economic history, and sometimes they may themselves interact unexpectedly to form really revolutionary restructuring on a new higher level of hierarchical structure.

All changes, however, do not involve misfit at the higher level. There are many pairs of structures that are so related topologically that it is possible to pass between them via continuous distortion without upsetting the overall connectivity. Liquids or solutions are of this type. Particularly interesting is the critical point where a liquid and a gas merge into each other—but only at a specific combination of composition, temperature, and pressure. Partially miscible liquids that differ in composition can do the same, for the solubility of each in the other increases with temperature and (unless something else like vaporization sets in) they will eventually merge. In crystalline solids such merging can occur between isomorphous solid solutions in which atoms are substitutable for each other without change of the lattice except in size. Diffusion, made possible by the random mobility of vacancies (vacant lattice sites), enables the components to become randomly distributed at high temperatures, but at lower temperatures the system has fewer thermal quanta to be incorporated, and the different atoms assert their preferences for like or unlike neighbors and begin to associate into either an alternating order on the scale of the unit lattice cell or larger regions of segregated clumping (figure 7). If the overall structure is tolerant, the interface between the structures can be diffuse and gradual, the interface energy being distributed over many units with the local strain never high enough to destroy topological continuity. The boundaries that can be drawn to divide such systems into discrete space-filling cells do not mark a disruption, but only a change in the second derivative of some property. The parts organize into an aggregation of cells of intermediate size each with exactly compensating positive and negative deviations from the average. These cells must, of course, ultimately fill space, but individually they do not necessarily have to be similar in size or to have any symmetry. Symmetry, indeed, has been grossly overemphasized in both art and science: its main value is in giving meaning to its absence, dissymmetry, without which there could be no hierarchy. A random aggregate of soap bubbles is a more generally applicable analogy than is a crystal lattice. Many structures come from the adjustment of boundaries and the internal rearrangement of groups of parts within an already existing structure. The shaping of valleys and mountains from the self-intensifying cooperative interaction between initial minute hints of a depression, the formation of cities and countries into cultural regions, the development of social institutions that inspire loyalty or moral commitment, or the enjoyment of games or art—all of these form as a result of cumulative regional weak interaction and are quite different from patterns resulting from rigidly specified identical neighbor interlock. They are a consequence of a long-range search for community reinforcement, though they only form after the overriding short-range demand has been satisfied.

On Art

The importance of the cumulative nuance as distinct from a brutally clear and simple statement is what much of art is about. Socially and culturally, the aggregation of nuances in diverse utilitarian things, daily custom and ritual gives richness without disruption, for the forces are gentle, not overly specific, and nearly invisible. Since the outcome is statistically determined and requires no basic change of connectivity while forming local clumps, they form easily and everywhere and do not require the disordered sites for nucleation of a first-order change (see the gradient at the left in figure 2).

In art generally the hierarchy rarely arises from such well-defined "crystallinity" as in figure 3, which was chosen to illustrate the analogy with the chemical phase. The requirement is simply that parts of some kind be perceptible and relatable to each other by the operations of the mind as the eye scans and compares, noting connections, invariances, symmetries, deletions, and modifications while mentally changing scale and orientation, and relating the current sensual input to other forms and concepts from recent experience or from more deeply embedded memory. Aesthetic pleasure seems to come as a kind of moiré pattern emerging on a higher level from the superposition of sensed and remembered images, somewhat as the experience of the third dimension arises from binocular vision. It is what is left over when what is expected has been canceled out. It seems always to involve some interaction between what is immediately visible and features on scales both above and below this.

Beauty can be perceived in a simple line (except a hard straight one) and it seems to arise from the relationships that the eye detects between the curvatures of different parts of it and the variety in less obvious more than one-dimensional qualities such as width, density, and sharpness of definition. Curvature can be uniform throughout as in a circle, or smoothly varying as in other conic sections, or in most natural forms where surface tension and viscosity serve to reconcile the geometric requirements of locally effective forces with distant larger ones. Other interesting lines may be complex, with reversals of curvature or with abrupt divisions into segments. The eye is repulsed by complexity if no order is detected, but it can be delighted by repetition, translation rotation, reflection, magnification, and other simple variations of the parts. As more levels of hierarchy can be constructed from the simple initial components the richer becomes the experience.

The retina of the eye receives information in two dimensions, but the mind interprets it in three or more. Features in the object in two or three dimensions can be suggested by outline or by internal shading, or both. Areas of pure color derive their effect from an unresolved but real substructure of wave-like photons producing a selective texture in the array of three types of color-sensitive rods and cones in the retina, much as the response to texture in a surface without color arises in just-resolvable granulation of the primary image. Some of the more enjoyable surfaces (for example, the grain of a fine mahogany table top or a Japanese sword) have an interplay between pattern and texture which, though two dimensional, suggests the unseen internal three-dimensional array.

As every painter knows, gradients of color or shading can distinguish areas without exactly defining them, and juxtaposition or repetition of such features are as effective visually in suggesting hierarchical relationships as are well-defined ones—indeed the very fuzziness invites the mind to search the field, the memory, and the imagination for similar but not exactly matching patterns and experience.

The style of a painting or other object lies in the way in which its parts (both physical and iconographic) are related to each other with repetitions or symmetries and contrasts that give a sense of the whole object presently in view and suggest a larger structural and cultural wholeness. The object itself is doubtless content to exist as a mass of atoms terminating in an external irregular surface, but the human mind, which learned in infancy to deduce a three-dimensional world from the changes in time of two-dimensional visual images and the sense of touch, notes similarities and differences between the various curvatures, junctions, shadings, and textures and then operates upon them to generate from the inconsistencies other levels in the hierarchy of matter and space and to extend into quite different kinds of worlds. The object, as it is experienced by a human being, is just one level in a hierarchical set, at the point where a series of crystalline and molecular and atomic structures reaching from below merge into a set of conceptual and cultural structures stretching both up and down.

Notice how all these factors contribute to the beauty of a piece of jade (figure 12). Overall, there is the approximate gross symmetry, but the many departures from the shape of the plane-surfaced Euclidian polyhedron that was first sawed from the boulder of nephrite give richness by inviting the detection of differences between one part of the outline or surface and another. The subtle curvatures arising from the soft-backed abrasive finishing techniques cause the sharp lines of junction to fair beautifully into more gently curved surfaces. Though basically simple and geometric, all of the features require and suggest something of a higher dimension. To this are added the local fuzzily-defined regions of shadings and color within the stone itself which add to the geometric relational possibilities, the slightly matte finish that gently disperses and softens the reflection of light from the surface, and the transparency and granularity of the stone which gives a deep three-dimensional sense of texture leading inward with decreasing clarity but no limit, while also recording something of the history of the earth and (via the deduced technology of manufacture and ceremonial purpose of the blade) something of man's history and aspirations. Every one of these features, by provoking the eye, the imagination, and the memory to find some resonances, facilitates the establishment of a pattern of interlocking connectivity like that between atoms making crystals but in a richer, less geometric mode, and characterizing many more levels of the hierarchy of existence.

As David Pye has shown in his fine little book *The Nature and Art of Workmanship* (Cambridge, 1968), the resonances arising in workmanship are often very subtle. The fact that the material itself guides the tool differently in different processes of working introduces changes in the overall relationship of curvatures. The smooth curves of surfaces approaching the edge of a jade axe that come about from innumerable abrasive particles moving against a slightly yielding and mechanically unconstrained backing would seem incongruous if other surfaces or outlines were present that had come from cleavage or from the geometric motions of a machine. These could be produced easily enough, but the eye would not establish larger resonances among them.

In the above discussion, I have dealt mainly with the nature of things that, like inorganic matter, are composed of parts that are rather simply aggregative, possessing some degree of interchangeability despite their complexity. A biologist would be more concerned with organisms in which a tree-like branching linear interconnection of specialized parts is superimposed on problems of cellular space-filling. In complex biological organisms as in social institutions, the differences between the parts become of importance comparable to the similarities, and integration becomes more a matter of functional interdependence than of simple structural interlock. There is a recognizable center of communication. The fast, almost volumetric communication via elastic strain in a crystal aggregate is replaced in biological organisms by the slower and less directed but more specific communication resulting from the broadcast diffusion of molecular units and the directed impulses of a nerve cell. Communication involves both emanation and absorption, and complex messages can pass only between complex structures. In both crystals and organisms resonant feedback is necessary, but the stability of the organism depends on different and complex communications between specialized parts. Without internal diversity and functional communication it would simply be an aggregate. One should note, however, that the functional interdependence is hierarchical, defining an entity by the interaction of its parts, differing from that in an aggregate mainly in distance and specificity. The units that become organisms are more individual, more polarized, and the boundaries defining them more distinct than in inorganic aggregates.

In a society, the specialized activities of hunting, ceremony, family life, cooking, or the production of food, tools or art represent different aspects of the individuality of their practitioners and products, but they nevertheless are interdependent within the culture, and each will develop in time some characteristic that marks the culture as well as the particular purpose. In the archaeological record, the style thus arising can most easily be seen in objects that are duplicated by mass production (not necessarily using machines) and are accepted in the environment of most members of the culture. In the fine arts a work is too individual to be a style until its essential aspects have been replicated, and new forms will be subversive until they have found enough copiers and patrons to be able to break through the conservatism of the old institutions and set their own style. An aggregate in which order and diversity are combined necessarily imposes differing status on its parts. In society there are two kinds of elite: the conservative managers involved in the functioning of an existing hierarchy and the radical intellectuals who are at the apex of a new structure that is about to develop. Neither is a common man. Neither can work by logical rote for both need aesthetic intuition.

Though my illustrative diagrams (as my own experience) have concerned structures familiar to the solid-state physicist and metallurgist, my examination of their hierarchical nature has led me away from science towards art, in which I have no professionalism whatever. While science has traditionally looked as exactly as possible at one level at a time in relative isolation, art seems to be in its very essence hierarchical. The viewing of art involves both the immediate recognition of simple patterns of color and resonance and symmetry in the parts, the slower recognition of less exact and conflicting relationships and of similarities persisting through changes of scale, orientation, and content, and the still slower perception (or perhaps subliminal absorption without perception) of allusions and cultural overtones. It is likely to involve a number of shifts of gestalt and a slow modification and renewed appreciation of detail. The creative artist seeks new discoveries in unexplored areas or new viewpoints in old ones where he senses relationships that have remained hidden. He makes suggestions rather than demonstrates conclusions with precision. Depending on the state of his environment his work may nucleate a prompt revolution in seeing, it may simply cooperate with other factors to produce a gradual change, or it may fail to establish any external resonance whatever and be forgotten.

All structures need units and some means of resonant interaction between the units that affects their being. The world is the sum of all structural interactions, and in a way it knows no parts. Yet reacting within it are nucleons, photons, and atoms selectively associating with each other, and human minds endeavoring to evolve structures to match. By thought we can select a level and a means of probing it. In time we might be able to probe in succession its workings on all levels, but we can never hope to think or sense them simultaneously. Despite this we can learn something. The nature of both the parts available and the means of communication within a system may change, and the means of externally probing it may differ, but patterns of interaction seem to have some consistency. Structure, most easily understood when presented visually, has much of the character of a universal metaphor. A field of view comprising relatively clear interrelations between parts in the middle and progressively more indistinct but nevertheless essential interactions towards the fringes can be applied at almost any level in almost any medium. For different purposes, different levels and different components will be chosen.

Nothing can be understood without at least a simplified glance at levels that are above and below the one of major interest. In both science and art the center of the limited perception of the human mind can be placed anywhere. In science the boundary of concern has traditionally been sharply defined to exclude unknowns, while in art the boundary is fuzzy and undefined: it can be anywhere, nowhere, or everywhere. The future seems to lie with a more extensive science, but it will have to be a multilevel science that, eschewing mysticism but not metaphor, will be able to pass continuously with a controllable focus and precision into the field of art.

There is no absolute scale, but being human, man should not, even if he could, wish to avoid emphasis on things at his own scale. And it is a very interesting scale, being at about the point of maximum significant complexity based upon the properties of the chemical atom. But regardless of scale, the meaning of everything lies in interaction, in the cumulative and changing history of associations on many levels which change at different rates. The structure that exists at any given moment is funeous, it is the product and the record of past associations and interactions, and it is also the framework within which future changes must commence.

Change is always unpleasant at the level most involved. The bitter pill that causes it has to be sugar-coated, and it has to be small in order to be swallowed. This is why art on an individual scale has so often been the precursor of large technological change—and why art on a social scale, which we so desperately need today, is so hard to experiment with.

Acknowledgments

Discussions with John Cahn, David Hawkins, Philip Morrison, and Victor Weisskopf have helped to shape the structural view of the world presented here. The conference on structure arranged by my colleagues Heather Lechtman, John Cahn, and Arthur Steinberg to mark my "retirement" in 1968 provided the impetus to consolidate the less scientific aspects of the structural principles that began to develop in my mind with studies of microstructure and X-ray diffraction in the twenties: the relationships with art came after a search for records of early metallurgy had made me a frequenter of museums, especially the Asiatic section of the Victoria and Albert Museum, the Fogg Museum, the Freer Gallery, the Museum of Fine Arts, and the Metropolitan Museum of Art whose curatorial staffs have been both tolerant and helpful. Most recently, membership in the informal group of people in the Cambridge area who call themselves the Philomorphs has served to enliven for me the problem of structure.

Bibliography

It is impractical to give detailed references to the prior literature. In addition to D'Arcy W. Thompson's *Growth and Form* (Cambridge: At the University Press, 1917), which marks the beginning of broadly integrated structural thinking, the following (some of which are listed in the Bibliography) have proved particularly helpful:

Rudolf Arnheim, ed., *Entropy and Art* (Berkeley: Univ. of Calif. Press, 1971).

H. S. M. Coxeter, *Regular Polytopes* (New York: Macmillan Co., 1963).

Gyorgy Kepes, ed., *Structure in Art and Science* (New York: George Braziller, 1965).

Gyorgy Kepes, ed., *Module, Proportion, System, Rhythm* (New York: George Braziller, 1966).

Wolfgang Köhler, *Gestalt Psychology* (New York: H. Liveright, 1929).

George Kubler, *The Shape of Time* (New Haven: Yale Univ. Press, 1962).

Arthur Loeb, *Space Structures, Their Harmony and Counterpoint* (Reading, Mass.: Addison-Wesley, 1976).

Howard H. Pattee, ed., *Hierarchy Theory: the Challenge of Complex Systems* (New York: George Braziller, 1973).

Herbert A. Simon, "The Architecture of Complexity," *Proceedings of the American Philosophical Society*, 106 (1962): 467–482.

Peter S. Stevens, *Patterns in Nature* (Boston: Little, Brown, 1974).

Paul A. Weiss, *Life, Order, and Understanding* (Austin: Univ. of Texas Press, 1970).

Hermann Weyl, *Symmetry* (Princeton: Princeton Univ. Press, 1952).

Lancelot L. Whyte, *Accent on Form* (New York: Harper & Row, 1954).

Lancelot L. Whyte, *Aspects of Form* (London: Humphries, 1951).

Lancelot L. Whyte, A. Wilson, and D. Wilson, eds., *Hierarchical Structures* (New York: American Elsevier, 1969). Contains a good bibliographical essay.

The many types of microstructure observed in polycrystalline materials are summarized by C. S. Smith in "Some elementary principles of polycrystalline microstructure," *Metallurgical Reviews* 2 (1964): 1–68, and illustrated in M. Bever, ed. *Metals Handbook, Volume 8, Metallography, Structure and Phase Diagrams* (Metals Park, OH, 1973). The role of aesthetic curiosity in the history of technology is discussed by C. S. Smith in three papers: "Art, technology, and science: notes on their historical interaction" in Duane Roller, ed. *Perspectives in the History of Science and Technology* (Norman: Univ. of Oklahoma Press, 1971), pp. 129–165, (preprinted in *Technology and Culture* 11 (1970): 493–549); "Metallurgical footnotes to the history of art," *Proc. American Philosophical Society* 116 (1972) : 97–135; and "Metallurgy as a human experience," *Metallurgical Transactions* 6A (1975): 603–623, (reprinted in 1977 as a separate booklet by The Metallurgical Society of AIME, New York, and The American Society for Metals, Metals Park, Ohio).

It will be clear that the author believes the nature of hierarchy to be most easily accessible to human understanding via visual patterns and material structures. References to mathematical treatments of group theory and algebraic topology (in which views parallelling those given in this paper are doubtless to be found) are therefore omitted. For similar reasons, the copious literature on systems analysis and that on structuralism in anthropology and linguistics is not cited despite some similarities in both problems and solutions.

Symmetry is a frequently cited aesthetic phenomenon in nature and science. Broken symmetry is discussed less often but has equally deep aesthetic properties. Philip Morrison demonstrates broken symmetries to be crucial in the process and realization of both art and science. "What we regard as highly satisfying works of art, even many natural things of beauty, contain broken symmetries. The symmetry is made manifest in some form, yet it is not carried out to perfection. The contrast, making visible both sides of the act of becoming, demands appreciation."

Broken symmetry is demonstrated to be crucial in the process and realization of both art and science. This theme is richly illustrated by examples of broken symmetry in particle physics, Cyril Smith's study of patterns of asymmetry in crystal structure, examples from architecture, and a theme from a story by Borges. We see that broken symmetry is the principle of dynamics and change. "No real example of an unbroken translational symmetry can exist because . . . our experience is not infinite. So you must come at last to an end . . . symmetry is broken by that end. . . . I suspect we react to the fundamental thermodynamic quality: an expression of symmetry, yet one not allowed to dominate exclusively, just as it cannot in the real world. . . ."

J.W.

ON BROKEN SYMMETRIES

PHILIP
MORRISON

The Idea of Indiscernibility

I touch on my subject with diffidence because it is so large a subject; I can introduce it only in a very sketchy and incomplete presentation. What I hope to do in pursuance of an aesthetic perspective is to suggest metaphors and connections. These are less than explicit, and probably less than entailed logically, but they seem to me to exist deeply in what we notice in the world. Much of what I have to say will be familiar to students of physics; less familiar to others. I make little claim for the novelty of these remarks. They belong in any insightful survey of our experiences in art or in science, most particularly in visual experience; among visual experiences, most particularly in consideration of architecture and industrial design because those products must work as well as be—a condition not imposed on other works of art. Function requires that the building at least stand up, and usually it ought to keep out the rain: a requirement very close to physics. I would by no means like to restrict my remarks to architecture. What I have to say will be of relevance to people concerned with the understanding of art in any domain of human action, from history to metallurgy.

To define the idea of symmetry is certainly not simple. I shall not try to make an all-encompassing or precise definition, against which a contradiction can quickly be brought, but rather to put out a kind of statistical approximation and then to enrich it by example. The idea is still alive and growing. We don't know all that the concept implies. I like best the idea of the seventeenth century philosopher Leibniz, who talked a great deal about symmetry not much employing the word, using rather what for him was a more epistemological word, a term at the heart of the matter. For Leibniz, symmetry is related to the indiscernibility of differences. Once you walk into the hall of a Palladian building, you can't quite remember whether you turned left or right. They look just the same inside. The two wings are indiscernible. They become discernible, one from the other, only when you understand that one is the mirror image of the other. But if I place the building up to a mirror, then the one becomes the other; once again the distinction is indiscernible in that sense. You see, indiscernibility stresses the idea of perception, which is why I want to use it as part of the definition. What is symmetrical under one aspect of perception may not be so under another. Symmetry has subjective quality. If I am color blind, I cannot tell the red-marked side of the boat from the green, the port from the starboard. The boat is, for me, perfectly symmetrical. I may say that the symmetry of the boat is fully present. But once I can see color, the left-right symmetry of the boat is broken by the two distinct colors, and therefore—most important—the pilot who approaches by night doesn't hit it. He can turn the correct way to avoid collision. It is no trivial matter: one is required to destroy the symmetry, the actual symmetry of approach. Without the lights you cannot tell whether the boat is approaching or receding. So naturally we mark the two sides to make them discernible. Discernibility depends on the channels of perception you use. Nobody could deny that red paint is inherently different from green paint: it contains a different chemical pigment, you can buy it from a different maker for a different sum of money, etc. Nevertheless, discernment does place some load on the observer.

We perceive the world by means of a complex system that we poorly understand—ourselves, of course: the eye-brain complex. Let me cite an indispensable aesthetic experience. It is the remarkable experiment presented to us only in the last decade by the work of Bela Julesz at the Bell Telephone Laboratories who presents to view two random dot patterns, one seen only by the left eye, one by the right eye. These present no figure, no structure. Each one is simply a set of five or ten thousand little black and white squares, like a much enlarged checkerboard. The squares are not colored in a regular pattern, but black and white at random. The right eye sees the dot pattern. The left eye sees a subtly different pattern, which indeed looks quite the same at first glance, a textured field of black and white. Within the left picture, however, a central square portion of the right picture has been inserted, exactly as it is on the right, except shifted a few columns to the left. The rest of the field is merely repeated. What you have then is a random dot field for each eye, except that a certain portion has been shifted *in toto*, maybe one thousand or so of the five thousand dots. When you view these two fields, one with each eye as is usual for inducing the stereoscopic effect, you gradually acquire a new and unexpected perception. The key word is *gradually*. Sometimes it takes seconds, or even tens of seconds, for the experience to form. You simply stare at the pictures without conscious effort of any sort. Something internal and undetected is going on within you, for after the delay you will see the square of dots, which was repeated in shifted form, float in space, a random texture above the random texture of the surrounding dots. Exchange the pictures between eyes, and the central square will float below the field of the others, farther away. It is a genuine and striking presentation of a binocular clue to depth, without any defined edge or familiar form.

The extraordinary thing is a sense of watching yourself because you're not conscious of any internal mechanism slowly coming to recognize the remarkable fact that the apparently random dot fields actually contain strong correlations. The little bit of difference in all these hundreds of pairs of dot distances is just enough to make the figure stand out in depth. Once you have that experience, you will never deny that there is some kind of internal, unsensed action. (It would be fascinating to develop some ability to perceive what's going on.) Something powerful is happening. After a few times the form comes more quickly, for a more complex structure like a figure eight or a bow knot, but it still takes many seconds for the depth judgment to jump out at you.

By extension, the idea of perceptual indiscernibility and therefore of discernibility, which is, of course, the other side, is most important. It has a great deal to do with our fundamental aesthetic perception of symmetries of all kinds and our delight in them. It is probably built into various symmetrical, "indiscernible" features of our mental processing: the logical circuitry, compiler, assembler, and all such tricks, in the metaphorical language of the computers. Of course, we know that constant scanning of the eye axis seeking lines and all kinds of other features is built in. Only slowly do we come to understand in any real way what is going on within. Be that as it may, I think it no miracle that human sensibilities are attracted to symmetrical presentations, for symmetry plays a major role in the world of which we are a part, of whose parts we are made. This is no philosophical mystery. Only if you see man as imbedded into a world of which he is no natural part, do you think that very strange. The relations of mathematics are exemplified as much in ourselves as in the rest of the world.

The Symmetry of Identical Modules

The first kind of symmetry I want to talk about in detail I must still touch delicately. For although it's most important, it has nothing directly to do with perception on the level of works of art because it is at a scale too small for us to perceive directly. Like the unconscious computer which we carry around, it lies so deeply within experience that it determines that experience in many ways. I'll try to adduce one or two of them, although we never directly perceive them at all.

The principal, most important symmetry of the world, in the sense of Leibniz, the chief indiscernibility, is the fact that the world is modular. I mean this: the particles that make up all our world are supplied to us in a few models, but in incredible numbers, all of each type identically the same. You cannot distinguish any single electron from another. Upon enlarging a little bit, you can, because you can measure its axial spin, but then you can only distinguish two classes of electrons: those with one hand of spin and those with the opposite spin. Among all with a given spin, you can't distinguish any one from its fellows. Indeed, you can keep one apart very carefully; for instance, put it in a box, and make sure it stays in the box. If you never allow it any chance of getting mixed up with another one, then, of course, you'll never lose track. Call the one in the box Joe, and you'll keep him straight! But that is a trivial victory because you might as well have called him Bill. It makes no difference. If ever you allow him to get into position where you might exchange two, where you take your eye away for a moment, you won't be able to tell the two electrons apart. There is no tag, no marker, nothing at all that we know to distinguish one electron from another, or one proton from another, or one hydrogen atom from another. Allow the possibility of a spectrum of states in which each of these structures can be found; in the case of the electron-proton it would be two states, in the case of the hydrogen atom maybe an infinite number, but all states with labels. I can say, "Yes, this hydrogen atom is in state number seven," and then I won't confuse it with any other which is in state number five, but all the *fives* and all the *sevens* are enormous populations of identical objects within which I can never distinguish.

That is the most profound symmetry in the entire canon of symmetries. It lies at the heart of our world of matter and radiation. It is this, and only this, which guarantees that gold is gold, and glitter only glitter.

I will discuss sketchily the physics of modularity, the treatment of identical particles. I do so because it seems to me that in part, at least, this must lie at the bottom of what we regard as the visual symmetries, the symmetries of perception. Let me describe it in the simplest possible way.

How can we impartially label three electrons? We begin by writing down the obvious labels, e_1:e_2:e_3, in an arbitrary order. Now we rearrange the three labels in all possible ways. One can choose any of the three numbers (1, 2, 3) to mark the first electron; for each such choice I have two choices among the two digits that remain, but no more, for after choosing two digits only one digit remains. For this example, then, there are six and only six orders for the three particles. Whenever I refer to the electrons, I must consider all six orders, and take only those descriptions of measurement which do not discriminate orderings. For example, the mass of one electron could be written as ⅓ the sum of the mass of each electron: $m = \frac{1}{3} (m_1 + m_2 + m_3)$. It clearly makes no difference in what order I choose the electrons; the mass so defined is impartial among the labels. I would calculate averages by taking into account equally all six orders of the three particles taken together. Then my theory is blind to labels as it must be.

This is no lecture in the mathematics of groups, but the obligation to deal with the world's marvelous modularity, plunging description of nature into an extraordinary metaphor, a very "mystic rose" of physics.

The rather dull task of labeling among identical objects at once describes something apparently entirely distinct: the geometrical transformations of a magnificently simple geometrical figure in ordinary space. That connection unites the linguistic and the visual, the algebraic and the geometrical, perhaps one might say, the left brain hemisphere and the right one.

Take an equilateral triangle like a capital Greek delta, Δ. Label the apex 1, and continue labeling its corners clockwise with 2 and 3. Now rotate the triangle about an axis through its center until it is indiscernibly changed. The peak has rotated one-third of a turn clockwise? Then you have changed 3 1 2 into 2 3 1. Once more by a third of a turn: 1 2 3. Continue as you will, always rotating the triangle this way, by one-third of a turn or any integer multiple whatever. You cannot rotate a triangle of this sort into any of its indiscernible, symmetrical presentations not described by the three orders we wrote down. Put them down even more schematically, as mere triplets of numbers in order: 123, 312, 231. There are no more results of rotation. But what of the six orders of the three digits? Indeed, here lies dizzy depth. There is a symmetry richer than rotation in the plane; it goes beyond plane rotation. It is mirroring, reflection, which turns a triangle into itself again, preserving its full indiscernibility. This can be thought of as placing a mirror against the paper, held along the right margin. Then the triangle image is indiscernible from the real triangle apart from the labels we artificially attached. But now they run 2 1 3 (never mind that the digits are unfamiliar in the mirror; we can still make them out). That order was not contained in the list of products of rotation. Now rotate the mirror triangle: we have the order of labels 132, 213, 321. These are all restricted to the results of at least one reflection. But taken all together, the results of pure rotation and reflection plus rotation yield all six orders, all the possible indiscernible positions of the delta, and exactly the results of label exchange. There are no more results possible, as there are no more operations which leave a triangle congruent with itself in Euclidean space: only rotations and reflections, and their combinations. The purely numerical exchanges gave the same results: the relation is called an isomorphism, and it lies at the heart of the theory of groups of symmetries. That theory is both rich and profound; here I have touched only the first lines of any text on the topic.

It is striking to see so deep a connection between Euclidean geometry and particle labels, which do not at all look geometrical. That is what we see on a large scale. If it were not true that one sample of stone is much like another, you couldn't have the sense of symmetry that one pillar is like a second pillar, and like a third pillar . . . in any temple colonnade. I believe at depth that is the same relation. It is a tremendous jump from indiscernible electrons to indiscernible limestone pillars, but it is no stupid jump; if art tries hard to remain pure and precise, cold and exact, it recovers more and more the hidden modularity of the physical world. One needs to carve all the pillars out of the same quarry—you don't want to take one black pillar and one white marble pillar if you want them to look the same—you'd then be emphasizing something different, indeed, you'd be breaking the symmetry, in a striking way. When the Shah Jehan built the Taj Mahal in his wife's memory, he intended to build another mausoleum for himself, much like the one that's on the bank of Jumna now, but across the river, in black instead of white stone, and connected to the tomb of his Queen by a golden bridge! His heirs wouldn't allow him to do it; they put him in jail first. (They didn't have the money, the fate of many magnificent projects!)

I argue that here is a simple but deep level in which we see symmetry arise in the inner nature of matter. Resembling it, mathematically one would say isomorphic to it, we had a relationship in Euclidean 2-space. Take those two examples together, generalize to other dimensions, and you have most of what the world calls symmetry. (So far we have talked little about symmetry *breaking*.)

By the way, to do this with a square, or to talk of *four* identical particles, is much harder. In the first place, there are 24 operations: $4 \times 3 \times 2 \times 1$. They are not isomorphic anymore to rigid geometrical rotation. You have to include the operations upon 3 and upon 2. The simplicity is gone, but still they contain all of the operations on a square, more than you can sort out. By the time you get to 12 particles, probably nobody's ever done it. The permutations grow to be bizarre.

It is remarkable—this is a digression I cannot resist. The physicist has demonstrated that the isomorphism of geometrical and permutation operations is not simple except in special cases. The only schemes that represent all numbers of particles are the very crudest schemes, of the following trivial kind: represent every operation by $+1$. If I perform two I get $+1$. No matter what I do, I get $+1$. So if you call every exchange $+1$ you require all permutations to produce no effect at all. That is said to be the symmetric representation. It is of course not one-to-one, therefore they say it is not "true." Nevertheless it is contained in the representations of all numbers of permutations. The other crude possibility is $(-1)^n$, which flips you back and forth between two classes, either the even or the odd value of n. These two are the only ways to represent all of the operations of permutation that can persist, not only for 3 electrons, or 4 electrons, 5 electrons . . . 10^{23} electrons, but any number you happen to have in mind. Now electrons have to live up to one fixed rearrangement of their labeling no matter how many partners with which you require them to rearrange. Otherwise we would experimentally see that once you put many electrons in a box, their symmetry would change. The world is not that way at all. We can conclude that only $(+1)^n$ or $(-1)^n$ are possible descriptions of what happens when you mix labels for any number of electrons. It turns out that relativistic quantum mechanics can show that it must be $(-1)^n$ for electrons, protons, and neutrons, and so on. But it is $(+1)^n$ for mesons and photons; only these two families of particles form the world: a profound proposition. We know of no breaking of this grand class of symmetries. Perhaps here I have said enough to enable you to see a little of how it must be on the level of fundamental particles. Among them we find structures of more complicated symmetry only when we arrange them in space. In the periodic table of atoms, or in the tables of nuclei, we are dealing with just such structures, geometry fused with permutations. I shall not build it up, but you have read about the electron shells, the intricate if approximate models which have some quality of truth. Chemists well know that spatial symmetry is found on the level of molecular bonds because modular symmetry exists on the level of the electrons themselves.

Symmetries of Space and Time

We also have more geometrical symmetries. Perhaps the most important of these is not rotation or mirroring already alluded to, but translation. That is, the fundamental symmetry of a set of columns of the Parthenon, or to invoke less lofty matters, the supports of an elevated expressway. There is symmetry in the sense of an unending iteration in length. You can use two dimensions of three dimensions. Crystals, as everyone knows, reflect this arrangement. Of course, they need not be as simpleminded as that, but yet they do require three unit directions and lengthy iteration of the translation. Notice, however, that such symmetries are necessarily broken in the real world. No real example of an unbroken translational symmetry can exist because the world is not infinite, or least, our experience is not infinite. So you must come at last to an end. A very long colonnade becomes a kind of a nightmare. You never come to the end of it. If instead there is an end to be seen, symmetry is broken by that end. The end is implicit in the finiteness of the structure, probably an important part of the aesthetics of such forms. If you're willing to make the form circular, rotational in nature, you don't have to come to an end because you go back upon yourself, but strict translational symmetry breaks at its inevitable end.

The subtle fabulist from Buenos Aires, Jorge Luis Borges, has told of a mysterious place where columns were built throughout a vast park by some mad prince; I suppose on top of each was a fine capital. They were beautiful, each a scarlet red. The next one you see as you ride along is just like the first. And you pass a dozen more, all about the same, as far as you can see, but if you ride a week you come finally to a white pillar! Yet each one has been like the first without any change. For, says he, the discernible difference between two hues of red that we call the same is, say, 1 percent along the scale of redness. Only after many pillars can you detect one colored a little different from the first. But you never see the two together; you see only a few neighbors, looking all the same. But by riding a week you come to a white pillar, without ever having detected any differences between neighboring examples. This tale is almost real. There is an analogy to the illusion in Escher's drawings, where monks walk up stairs all the time yet never get any higher. In both cases the paradox arises because the global and the local decisions of equality are not the same, a strong point for the idea of symmetry as discernibility.

Leave the fanciful things. Most familiar of all symmetries is human bilateralism, we are the same roughly speaking—our faces look the same, directly viewed or in a mirror. You and your image are one identifiable person. We know, in fact, this is always broken; it is rare to find someone for whom his good friends cannot distinguish a direct view from a mirror image. The flipped photographic print never looks comfortably familiar. It is hard to say what subtle differences allow one to tell. In the same way left hand and right hand are not exactly the same; the left half of the cerebrum and the right half are definitely not the same, one related to speech, the other to geometrical perception. Near the midline, the heart has its strong muscles over on one side. In all but 1 person in 10,000, that is the left side. On the other hand, we know that life reflects an inner molecular handedness; the amino acids in every protein, for example, are the so-called L-form, and their mirror enantiomorphs are not commonly found in nature, except in very special circumstances. You can actually feed bacteria on a symmetrical molecular mixture, to produce a handed residue, because the bacteria won't use up the wrong half of the foodstuff they're given. Here the symmetry, of course, is gone; the indiscernibility is removed for ancient historical reason. We don't really know how, but we have to say that the chain of life is linked, one form always using the molecular parts from another. You couldn't share in the chain of life if you didn't share this asymmetry. As far as we know, nothing whatever prevented life from starting the other way. Nothing would work differently. Had it started the other way, there'd be no necessary distinction except by confronting the two faces. That instead life sticks to one way predominantly is a sign only of its unity, its universal kinship, not a sign of special preference, except maybe at the beginning. Perhaps subtle forces caused the earliest life to choose one direction rather than another? It could have been chance, or it could have been the presence of particular handed crystals in one chance place where it all began. We don't know, it is a goal of active investigation.

There is a strong line of studying symmetry in the domain of fundamental particles, not only the enduring particles, electrons, protons, and so on, but the great zoo of unstable particles on which physicists are working. This is not my topic; I want simply to mention it. But there is, of course, a discernibility known now for more than fifteen years which I can express best in this way. Let me place a disk on an axis so that it can turn around freely, a real metal disk, some carefully-made piece of gyroscopic machinery. Now I paint the surface of this disk with a certain radioactive substance, for example, cobalt-56. If I made this disk well, placed it in a vacuum, and allowed the 50-day half-life of cobalt to pass, in order to allow most of it to decay, it should spit a lot of electrons outward. But inward the electrons find the disk rather thick, so they get caught and remain fixed in the disk. The disk loses preferentially those electrons which started outward. Such a disk will spin spontaneously in a clockwise direction. That's built in to the way that cobalt-56 decays! (The matter is complicated because each atom gives out two particles—neutrinos which are never caught and electrons which can be but always share in such a way that the higher-energy electrons preferentially take off one direction of spin; so do the neutrinos, but they're not getting caught!)

Such an experiment is not practical but is entirely conceivable! Less macroscopic analogies to it have been done with care. This was striking; it was anti-intuitive, and it won the Nobel Prize for Lee and Yang after Wu and others did the experiment in 1957. It was a wonderful result, if not very much more came of it. What was immediately shown was that, to be sure, you could distinguish the right hand from the left hand; there was a direction built into our world. Pauli said, "God was displayed to be a left-hander." But not so. It turns out that if I take other radioactive elements, the disk might not spin the same way. That's all right, one can always recognize cobalt-56 by counting the number of nuclear protons. It seems still perfectly unambiguous. But there's another experiment. Count the protons, everything's correct: I build the same disk, but lo and behold it spins the other way! Yet I've done exactly the same thing, save for one little fact. What is needed to make it spin the other way? A source of anticobalt-56. What is anticobalt-56? It is cobalt-56 in which the protons are replaced by antiprotons, and the neutrons replaced by antineutrons. You can't buy that. But we have made antideuterium. So it is only a question of care and effort; we might in principle be able to go up to cobalt-56. We don't doubt that it would work the same way. But it shows that discernibility is like the mirror. The left and the right hand are exactly the same. I don't know whether a mirror has been interposed or not. In other words, the mirror's image of the left hand is just like the right hand. In the same way, the antimatter image of cobalt-56 changes its handedness. So one does not know whether the sample is from the real world or the antiworld; again there is no way to distinguish left from right. There is no experiment which will enable me to distinguish anticobalt-56 from normal cobalt-56. If I bring the two together, they mutually react. Both disappear in a burst of unstable particles and radiation. It is generally believed that the matter-antimatter mirror is not exact. These are deep questions of broken symmetries, of fundamental interest to the particle physicists of today.

The idea of symmetry, the idea of the indiscernibility of transformations has been fairly straightforward. We change plus to minus, one direction to another, an object to its mirror image, rotation to translation. One profound result can be added. Recall the foundations of classical and, indeed, of quantum physics: energy, momentum, and angular momentum (like spin) are the three great conserved quantities of classical and quantum physics. Everyone knows energy is conserved, momentum too is conserved, and angular momentum is conserved in all processes of which we know. If you gain energy, some other system must have lost it. If you gain momentum, some other system must have lost it. If you gain angular momentum, some other system must have lost it. If you start with none, you can't get any energy at all, because it's just a number, without direction. If you start with no momentum, you can gain momentum, but only by sending away the necessary amount in the opposite direction. If you start with no spin you can gain spin, but only by sending out spin in the opposite direction. Those are the great constancies of physics. We know no exceptions to them; they hold very widely, and a very few others, only a few, like the constancy of electric charge, complement them. It is possible to show that under reasonable conditions the existence of the conservation of energy is essentially a statement that in the development of a system, one time is like another. The conservation of momentum is a statement that in the development of the behavior of a system, one position in space is like another. There's no difference; they are indiscernible. Angular momentum constancy says essentially that one axis direction is like any other. Physicists are used to saying that these are symmetries because there are constants during any process which can be shown inescapably to come with the space-time symmetry of any system. If I take *the whole* of a system, carefully allow nothing to be left out, it doesn't matter at what time I consider that entire system; it's going to work the same way. It's true that if I have an alarm clock, it'll wind down. But if I can deal with the whole system, the winding up and winding down, the effects of air, and so on, if I take all of it into account, I could have done the same thing a million years ago, and it would work the same way. Absolute time makes no difference. On that basis you can show that the abstract quantity we call energy must be conserved. In the same way, if absolute space makes no difference, momentum must be conserved. If absolute direction makes no difference, angular momentum or spin must be conserved. It is probable that all symmetries are a little bit broken because the real world does show some kind of structure in space and time and direction. We can't put our fingers on it yet, but the universe is not the same in all directions. It is more or less the same, but when I point *here*, I do not see the very same galaxy as when I point *there*. On the average, I see the same kind of thing, but when I point at the Andromeda Nebula, it is not the same as M33. Maybe that specificity shows up, but no one has found it yet. People have looked hard, down to a part out of 10^{15} and 10^{20}, and they haven't found it yet. But it might be there—a failure of the conservation of energy or a failure of the conservation of angular momentum.

Symmetry and the Act of Becoming

The next point lies closer to the surface. Examine the changes that go in the world, all kinds of changes: the growing of plants, the froth moving on the waves, the flash of a lightning bolt through space, how each of us ages. Try to give an account of change in some broad, general language. One thing that you might say is that there are certain principles of economy; each natural process spontaneously seeks a minimum of some kind. The philosophers of the eighteenth century put a great deal of weight on that. They thought that here was a great economical principle, Nature seeking to do great things with least means. We have a less broad view of it now because we recognize that these minima are not global, but local. That is to say, you might find a new way of doing something by going along a completely different path, still easier than the way it actually goes, but on a different branch of behavior. For example, you know how light goes from air to water; it breaks toward the normal. It's an easy demonstration: if I know the speed in air and the speed in water, the shortest path from start to finish is the path given by Snell's law. That was a grand principle in the old days because they felt something marvelous in the economy of nature. What this means is that the shortest path compared with all nearby paths is realized. But if I put a mirror in the way, I can get a much longer path, yet still just as natural. It turns out that they are often the *longest* paths that you can take. It doesn't matter to us; it's not the shortness that counts for us, it's the extremal nature. In the near neighborhood of the actual path, there are many neighboring paths almost the same length. Those are the paths that are realized; whereas any path that goes in a crazy excursion across the countryside has nothing comparable to it in its near neighborhood at all. Such paths never add up to anything. The real world works on this "most probable" scheme. We ought not to make a great deal out of minimal (or maximal) properties, but there are many situations where we imagine that a minimum state will prevail. For example, a soap bubble is spherical; clearly the spherical form gives you the minimum surface area for the given volume enclosed. If you ask what is the energy, there's the gas pressure inside the bubble and the surface energy of the film. Minimize the total energy by making a spherical bubble of a certain size for every given pressure. You can see how that would work. All I want to argue is that the fact that it is minimal gives a mathematical target at which to shoot. (Actually, it's only the extremal that really counts, the reinforcing property, near the minimum or sometimes near the maximum, that is important.)

Notice when I talk about soap bubbles, I am far from talking about the domain of atoms. When we think of soap bubbles, there is a volume inside full of gas and the filmy surface of the soap bubble. Of course, it's not a surface; the film, too, is a volume. In the atomic view of the world, there is a collection of certain atoms of high density in one place and others of low density in another; certain forces in one place, other forces in another; there is no distinction in kind. But it is a good approximation to deal as though one were in a true continuum containing various continuous substances; that's what we do. The film is different from the air. That is not a molecular view of the world, but a continuum view of the world, suitable for almost all work in classical physics, or on the scale of art we can perceive. You rarely have to consider molecular structure itself.

Let me describe a situation in which the breaking of symmetry during the act of becoming is made manifest to the eye. There is a wonderful, if expensive, toy which does the job remarkably well. (It is the invention and product of a clever designer in Montréal and is sold under the name *Atomix*. Many a reader will have enjoyed one.)

You see a transparent plastic slab about the size and shape of a paperback book: say five inches square and a little more than an inch thick. It is not, in fact, a solid slab, but a sandwich. The two pieces of "bread" are carefully fastened together with a sealed air space as the "filling" between, a very thin filling, little more than a millimeter thick, over a square somewhat smaller than the two defining plastic pieces of the sandwich. That air space is occupied by some five thousand small steel spheres, similar to the ball of the ball-point pen. They are good commercial spheres, quite alike, indiscernible to the eye. They can very well represent individual atoms as the designer intends. Of course, they are grossly inaccurate on the atomic scale; no two are alike to microscopic examination. You could learn to call each one by name, unlike atoms. Human scale does not permit the modularity of the particle world; after all, each little sphere contains more atoms of iron than the number of snowflakes which fall worldwide per year. They are not identical spheres, but they are indeed similar. They are smooth and free to roll within the air space, where they take up perhaps two-thirds the volume. The space they occupy is so thin that they remain spread out in a single layer however they move; there is no room for one to pile on another along the thickness of the air space.

What a mobile and wonderful layer the small spheres make! They respond to every tilt of the device by rolling downwards under gravity, until they touch the walls or other spheres and can no longer fall. They are spheres, and in the uniform packing which similar spheres can take up, each one touches six neighbors in the plane located symmetrically around the central one. They arrange themselves into such hexagonal patterns over the whole area of their space, obedient to the requirements of seeking a minimum energy. They can't pack more closely, for the steel is rigid and little deformed under the small weight. They fall until they touch, for the force of gravity overcomes the little friction, and they stop, rolling only when another sphere or the plastic wall matches the gravitational pull. They remain in place in their thousands once they have come to the equilibrium of minimum energy, in a striking crystalline array, a plane layer of identical spherical "atoms" under minimum energy, exactly analogous to the structure of crystals, but, of course, on a very different scale, one perceptible to the eye. No plans for design were needed; the orderly crystallinity, the symmetry we see, arises out of the automatic play of force on an array of identical spheres. The spheres are indiscernible; each one acts like its neighbors, and the enforced result is the hexagonal packing clearly visible whenever the device is viewed. Order has appeared fundamentally out of the symmetry of the sphere and the rule of modularity. No designer need intervene.

Yet that order, that symmetry, is not perfect. It is definitely broken, but not in a chaotic way. Here and there a hole occupies a point where a sphere would fit. In rolling there, some spheres by chance so fitted that beneath them a hole formed which could not be reached by a sphere that would have fitted there. A hair more shaking and that hole would have filled. As chance would have it, the place remained empty. In a typical shake, there are dozens of such spaces left. But its neighbors pretend the whole is filled and make their hexagon anyway. It is much better for the energy minimum that they each have five real neighbors than that they act somehow crazily. Rarely there are two holes nearby. More often there are lines of slip and mismatch—quite understandable and quite like those which the physicists have found in real atomic crystals. A whole line slipped out, and two neighbor lines made up the space by a jump of only one ball spacing. Or one region of five hundred balls arranged itself in the hexagon pattern while the balls rolled; so did a region nearby. But there was no way in which one region could signal the orientation of its hexagons. The two meet at a boundary line along which there have to be a few regular spaces. The texture looks different; this is what in real material is called crystal grain. All the most important defects of real crystals are exemplified; but that is not our story, helpful as it is to generalization.

No, the point is that the perfect symmetry is broken. It must be arrived at by local force interactions, but any chance event will modify it slightly. As the order continues on in space, sooner or later it must find some other center of order with its own modification. There a conflict will occur; the final order is partial, regional, incomplete. If with arbitrary slowness we allowed the balls to roll one by one, and avoided all interference, if we intervened to make sure that one hexagon pattern formed with the same orientation as its neighbors, we might make a perfect "crystal." (A patient and dexterous player can, in fact, do this with an *Atomix* model.) But in the real world of finite rates, of noisy interference, of blind forces of assembly and disassembly, perfection is, not inconceivable, but rare to the point of no practical significance. Large systems cannot be perfect unless they are formed at a snail's pace. Big gems grow slowly and are rare. Local symmetry and global repentence is all that a world of atoms can achieve. Symmetry almost always is broken by growth, as by destruction. Most often the system has to satisfy the minimal principle. But if it anywhere doesn't reach minimum, there's going to be a defect. The memory of a defect will force a defect somewhere else. The whole thing has got to work out in the long run.

The symmetry of the array is broken, really because of time. Since no process, certainly no spontaneous process, can go on for infinite time, some irregularity must occur to propagate its mismatch faintly throughout the entire system. It is just as wrong to expect that perfect symmetry to be present as it would be to neglect the symmetry entirely. The first sort of error would be to ignore the whole idea of the energy minimization principle. The other sort of error is to imagine that the whole system is constructed with infinite time available, or, if you like, with every perfection as a goal. But in reality structure is built by local forces. They do the best they can at every point, but they cannot achieve perfection. When once they fail the effect must be made up somewhere in the system as a whole. The spheres give up finding an absolute minimization at every point in favor of a practical solution. In most places the structure can come close to the minimum. With a finite rate of formation, it will do the best it can to adjust. It doesn't wait until somebody comes to make exactly the right geometry. It can't grow that way. It has to grow by local events. The global adjustment must be propagated, yet it can't propagate quickly enough to satisfy the minimal condition everywhere.

This is a metaphor, an analogy, for what happens in all real processes in the material world. We can never minimize energy truly in the presence of real processes that require real times of growth. Everything must grow at some rate. We always trade off error against time; we trade an impossible symmetry for the scarce but possible energy. We have to pay in the coin of energy to be absolved from a need to try hard for exact form. That is the nature of broken symmetries in the world. Order forms spontaneously all the time; the Second Law has nothing against that at all. The Second Law insists only that if you make order, you must pay for it. You must pay in energy. Of course, if you make an improbable arrangement which is yet very orderly, you have had to use some of the energy that otherwise you might have gained. The reason is quite clear: The world is immersed in disturbance. The world does not consist of isolated subsystems sheltered in their growth. When I shake the ball bearing model to try to make them move into place, that corresponds, if you will, to a sort of Brownian motion at a high temperature. If I went in there with tweezers to move each one, carefully enough so my hand doesn't slip, I might get them all arranged perfectly evenly, at a very great price. Any speed of working is going to breed errors; those errors must break the symmetry.

The world is governed by minimal principles which apply locally in every place. They cannot be expected to overcome the constraints imposed by time and topology. It is just as remarkable to see an absolutely perfect crystal as an assembly of the right atoms with no crystallinity at all. Neither could occur. The only way to have an absolutely perfect assemblage would be to have the whole thing formed at the absolute zero temperature—no randomness. But, of course, at true absolute zero the rates of formation would go to zero. Nothing would ever happen. Only an isolated world where nothing ever happens can be perfect. (Is that Nirvana?)

Broken Symmetry and Art

I am not justified in making a moralizing conclusion in aesthetics, but I can state a prejudice. . . .

What we regard as highly satisfying works of art, even many natural things of beauty, contain broken symmetries. The symmetry is made manifest in some form, yet it is not carried out to perfection. The contrast, making visible both sides of the act of becoming, demands appreciation. A soap bubble is beautiful. Somehow everyone would agree that it has a kind of simplicity, a coldness, which bars it from the category of great beauty. In fact, the very reflections and color changes which make it something other than a perfect sphere enhance its beauty. A cube of glass, too, is a beautiful object but no work of high art. If you see the work of a lapidary, a rough crystal, the crystallinity plain on some faces, but hidden in the matrix of others, it is a more satisfying object. I suspect we react to the fundamental thermodynamic quality: an expression of symmetry, yet one not allowed to dominate exclusively, just as it cannot in the real world, for some feature always breaks every macroscopic symmetry in the end.

Arthur Miller introduces in his essay a new approach to the development of quantum theory, emphasizing the drama and the force of imagery and metaphors in the development of a theory of atomic phenomena. Aesthetic judgments played a major role in this period—an aesthetic of waves and/or particles and the choice of a mathematical formalism. Whereas Heisenberg's mode of thinking committed him to continue to work with a corpuscular-based theory lacking visualization, Bohr, Born, and Schrödinger believed otherwise; their need for the customary intuition linked with visualization was strong. Heisenberg's reply was that a new definition of intuition was necessary, linking it with the mathematical formalism of his new quantum mechanics. Visualization was regained through Bohr's personal aesthetic choice of the complementarity of wave and particle pictures, thereby linking physical theory with our experiences of the world of sensations. What is so remarkable about this period is that the intense struggle between these physicists surfaced in their scientific papers.

J. W.

VISUALIZATION LOST AND REGAINED: THE GENESIS OF THE QUANTUM THEORY IN THE PERIOD 1913–27

ARTHUR I. MILLER

Introduction

There is a domain of thinking where distinctions between conceptions in art and science become meaningless. For here is manifest the efficacy of visual thinking, and a criterion for selection between alternatives that resists reduction to logic and is best referred to as aesthetics. To demonstrate this domain we can examine a case study in the history of science—the genesis of quantum theory in the years 1913–27. The notions of aesthetics held by the dramatis personae of this period will not always be associated with the choice of a pleasing visualization, but sometimes with a choice of thema, such as continuity or discontinuity, with the choice of a particular mathematical framework, or with combinations of these criteria. And sometimes their aesthetic will change.

In 1929, Niels Bohr could write that the orderly progress of research culminating in a consistent theory of atomic phenomena was made possible because of the "conscious resignation of our usual demands for visualization."[1] Another pioneer of the quantum theory Wolfgang Pauli in 1955 hinted at "a brief period of spiritual and human confusion caused by a provisional restriction to *Anschaulichkeit*"[2] and ending in about 1929. The German *Anschaulichkeit* means visualization through pictures or mechanical models. It was used much less frequently in the German scientific literature of the 1920's than *Anschauung* for which there is no equivalent term in the English language. Unless specified otherwise, *Anschauung* will refer to the intuition through the pictures constructed from previous visualizations of physical processes in the world of perceptions; this best fits its intended meaning in the period 1925–27 by Bohr and Pauli, among others. For convenience I shall sometimes designate this kind of intuition with the term "customary intuition."[3]

A measure of how dramatic was the search for a consistent atomic theory, and of the ambiguities lurking at every turn, is the disagreement between the accounts of Bohr and Pauli. The path to the quantum theory of 1927 was not an orderly progression from visualizable models to a mathematical formalism whose description of matter and phenomena in the atomic domain defied visualization in the ordinary sense of this word. Rather the situation was closer to the one recalled by Werner Heisenberg where the physicists experienced despair and helplessness because of their loss of visualization and of their distrust in customary intuition. It was a period when such time-honored concepts as space, time, causality, substance, and the continuity of motion were separated painfully from their classical basis. The rejection, one by one, of the tenets of classical physics was distressing because classical physics had become so commonsensical, so reflective of the world in which we live. Since it was a causal theory, a particle's motion was continuous and its position could be predicted with, in principle, perfect accuracy. There was no reason to doubt that our intuition could also be extended to phenomena involving the microscopic electron. So it was hoped with confidence that the discontinuities which *emerged* from Max Planck's theory of 1900 for the continuous spectrum of the radiation emitted from a hot substance could be removed somehow in a more fine-grained theory. But in 1913, the twenty-eight-year-old Niels Bohr proposed a new kind of theory for the free atom. It was *predicated* openly upon discontinuities and gross violations of classical physics. Yet despite its transgressions Bohr's theory retained the pleasing picture of Rutherford's atom as a miniature Copernican system. But Bohr did not consider the theory complete because it conflicted with classical physics. However, by 1918, the break between Bohr's theory and the classical physics became more striking when Bohr folded into his theory a kind of mathematical description for the unvisualizable quantum jump that was forbidden by the accepted view of the classical physics of particles; namely, a priori probabilities which renounce from the start any knowledge of the cause for phenomena. By 1923, the limitations of the picture of the atom as a miniature solar system had become forcefully apparent. The basic problem was that it could not treat quantitatively atoms more complex than the miniature solar system with one planet—the hydrogen atom.

So atomic physics began to slip into an abyss where the planetary electron went from a tiny sphere to an unvisualizable entity. What is so remarkable is that this surrender of visualization was precipitated only in part by empirical data, that is, problems with complex atoms, but more as a result of Bohr's aesthetic choice between the opposing themata of continuity versus discontinuity. Bohr's decision was to reject the light quantum in order to retain the traditional dichotomy between the continuous radiation field and discrete matter. This limited duality was his aesthetic in 1923–25 and it was not linked to visualization. Heisenberg, at age twenty-three a contemporary of Pauli, spent late 1924 to early 1925 at Bohr's Institute for Theoretical Physics at Copenhagen; he was already making his mark in physics. Whereas for Bohr the loss of visualization was painful, Heisenberg found it to be congenial to his nonvisual mode of thinking. In late 1924, Heisenberg's analysis of the data of Wood and Ellett for the scattering of polarized light from mercury and sodium vapors convinced him and Bohr that the loss of visualization might very well be permanent and that since mechanical models had failed the existing mathematical formalism of quantum theory must be the guide. In mid-1925, Heisenberg formulated the new quantum mechanics based upon an unvisualizable particle and essential discontinuities and expressed in a mathematical formalism unfamiliar to most physicists. But the desire for visual thinking and for the intuition associated with the world of perceptions, or with classical geometrical concepts, manifested itself repeatedly in the papers of those most closely associated with the new quantum mechanics—Bohr, Heisenberg, and Max Born who, at age forty-three, was Bohr's counterpart at Göttingen. Of great concern to Heisenberg was that due to the loss of visualization and customary intuition the new theory was risking internal contradictions.

Even more remarkable than the loss of visualization was how it was regained in the period 1926–27 when empirical data played almost no role. Rather the path to regaining visualization is characterized by the high drama of the intense personal struggles among the dramatis personae over their choices of the themata in conflict—continuity versus discontinuity, the usefulness of mathematical models versus mechanistic-materialistic models, and whether to maintain causality. These are among the themata that have emerged from Gerald Holton's pioneering historical case studies as having been of great concern to scientists through the ages. Holton refers to them as "thema-antithema couples."[4] His studies reveal that a scientist's criteria for the choice between a thema-antithema couple cannot be reduced either to logic or to a suggestion derived directly from experimental data. Holton's observation and terminology are applicable here because the choice is based upon the individual scientist's aesthetic. What is so fascinating about the genesis of quantum theory is that not only does the personal nature of the struggle between themata emerge from the scientific papers of the period, but the themata clash here as never before in the history of science. The stakes were high in 1926–27 because the outcome could determine the course of physical theory.

A new figure enters in 1926. Erwin Schrödinger at forty-three was a Professor at Zurich and an outsider both geographically and in thinking from the small band of physicists concentrated mainly in Copenhagen and Göttingen who were almost entirely responsible for the development of the quantum mechanics. Schrödinger's sentiments were closer to those of the continuum-oriented physicists at Berlin, in particular Einstein and Planck. Schrödinger felt so "repelled" both by the lack of visualization and the mathematics used in the quantum mechanics that he formulated the wave mechanics. His aesthetic was linked to a continuum-based physics which visualized that atomic entities were composed of packets of waves. Nevertheless, he claimed that this was better than no picture at all.

Although Born had been godfather to the quantum mechanics, his desire for visualization led him to enlarge his aesthetic from the one held currently by Bohr and Heisenberg to a viewpoint in 1926 in which particles are guided by waves from Schrödinger's theory. Born considered these waves as transporting only probability. His theory permitted the use of pictures for scattering processes but distinguished between the notion of causality as it referred to scattering processes and to phenomena within the atom, for example, transitions.

Heisenberg, from the beginning, had found the wave mechanics "disgusting" and was further enraged at Born's desertion of the cause of quantum mechanics. This situation and conversations with Bohr in late 1926 reinforced both Heisenberg's aesthetic and his nonvisual modes of thinking, leading him to boldly *demarcate between* intuition and visualization. The results in early 1927 were the uncertainty relations and the rejection of the only causal law that this viewpoint could consider—classical causality.

To Heisenberg's consternation and eventual distress Bohr pressed him relentlessly not to publish his results. For Bohr had realized that only the complete wave-particle duality for matter and light could lead to a consistent interpretation of the quantum mechanics containing visualization. However, the causal laws could no longer be associated with space-time pictures, but with the conservation laws of energy and momentum. This viewpoint is embodied in Bohr's complementarity principle of 1927. The analysis leading to the complementarity principle plumbed the very depths of knowledge down to the formation of ideas themselves. Bohr emphasized that the necessary prerequisite to this analysis was the discovery that visual thinking preceded verbal thinking, and linked to visual thinking could only be the aesthetic of the symmetry offered by the complete wave-particle duality.

The roots of complementarity are as varied and deep as its ramifications into other disciplines and so I believe it is apropos here to mention Bohr's lifelong interest in art, especially in cubism. The Danish artist Mogens Anderson, a friend of Bohr, recollected what most impressed Bohr in cubism: " . . . face and limbs depicted simultaneously from several angles. . . . That an object could be several things, could change, could be seen as a face, a limb and a fruitbowl."[5] We shall see that this motif has striking parallels to that offered by complementarity where the atomic entity has two sides—wave and particle—and depending on how you look at it (that is, what experimental arrangement is used), that is what it is.

So in late 1927, Bohr raised physics from the darkness of the abyss and the drama was completed. We are reminded here of one of Bohr's favorite sayings from Schiller:

Only fullness leads to clarity
And truth lies in the abyss.

Visualization Lost (1923–25)

By the first decade of the twentieth century it was taken for granted that physical theory should be a reflection of the continuity that we observe in the world of our perceptions. Lack of continuity would mean breaking the connection between cause and effect, that is, violating the law of causality. The law of causality in classical physics states that given a particle's initial position and momentum, its continuous path can be predicted with absolute accuracy, at least in principle. The classical physics is also deterministic because there is no reason for a particle to be unable to occupy any point on the path deduced from the equations of motion; equivalently, since the continuity of the equations of motion ensures determinism, there is no difference in classical physics between causality and determinism. Given the particle's initial momentum and energy, the values of these quantities can be determined for any point on the particle's path from the conservation laws for momentum and energy; in classical physics these laws are linked to the picture of a particle moving through space.

The newly discovered electron was visualized as having a definite shape, and when accelerated it becomes the source of electromagnetic waves (light) spreading out like the circular ripples caused by a stone thrown into water. This dichotomy between discrete matter and the continuous or wave picture of light was acceptable in classical physics. Even though the electron introduced a discontinuity into the substratum of nature, the laws governing physical theory were required to be continuous.

But in 1900, Max Planck discovered, to his horror, that he might have opened a Pandora's box.[6] The issue concerned his radiation law describing the continuous spectrum of the light emitted from a cavity within a hot substance. This is a most interesting spectrum because it is independent of the substance's constitution; such phenomena fascinate physicists and until 1900 a formula describing this spectrum had eluded them, despite intense efforts. Since the spectrum is independent of the constitution of the cavity walls, it matters not at all what model is used for the source of radiation. Planck decided to use a currently accepted model that is both easily visualizable and whose equations of motion are easily soluble; namely, the electrons comprising the cavity walls are charged spheres on springs or charged oscillators. According to Planck's theory of 1900 the continuous spectrum of radiation from the cavity results from the charged oscillators emitting and absorbing light energy discontinuously in bursts or energy quanta. This gross violation of continuity caused Planck to devote many years of his life seeking other theories for the radiation law consistent with classical physics.

Planck's reputation notwithstanding, his theory of 1900 was politely ignored until it was resurrected twice over by an unknown patent clerk third class in Bern. Albert Einstein proposed in 1905 that certain phenomena could be described easily if light were considered as having a granular structure, thereby flying through space like a hail of shot or light quanta.[7] Sensitive to the mood of the times, Einstein in 1905 did not use Planck's radiation law in calculations supporting this viewpoint. Nevertheless, it is clear from the light quantum paper of 1905 and his later writings that he knew of it and that he was boldly extending its consequences beyond the quantization of energy. Among the critics of the hypothesis of light quanta was Planck who could not construct a visualizable model in which light quanta produced interference,[8] whereas almost three hundred years previously Christian Huygens had proposed a visualizable construction for light waves interfering. Einstein in 1907 used Planck's law of radiation (without light quanta) toward a quantum theory of solids.[9] This research was discussed at the first summit meeting of physics—the Solvay Congress of 1911. The theme was the theory of radiation and Planck's energy quanta; for by this time the special relativity theory, another product of Einstein's Annus Mirabilis of 1905, was considered as well understood. This Congress provides us with a glimpse into the accepted view of what a physical theory should be.

Perhaps the best survey of the Solvay Congress was written by Henri Poincaré,[10] who placed a high premium upon visual thinking and those physical theories whose mathematical framework exhibited the highest degree of symmetry.[11] Poincaré focused upon his recent proof that a physically meaningful description of cavity radiation can be obtained if and only if the energies of the charged oscillators were restricted. This result deeply disturbed Poincaré because it meant that any law for cavity radiation, for example, Planck's, must possess discontinuities. But, Poincaré lamented, should "discontinuity reign over the physical universe"[12] then determinism would be invalid since all positions of the charged oscillators would not be permitted (the charged oscillator's energy is dependent upon how much the spring is extended from its equilibrium position). Fundamental changes in the laws of mechanics would be necessary because they could no longer be considered as describing continuous motion. Poincaré, as did most others including Planck, preferred a wait-and-see attitude. With confidence they believed in the possibility of extending our customary intuition into the domain of the atom, and of finding there the mechanism or cause for the emission of radiation by the atoms lining the cavity wall; as expected there would be a continuous causal chain. The search for causes, with the optimism that they would be found, was a hallmark of classical physics; for this was necessary in order to maintain both causality and visualization of physical processes.

Should this search fail a dramatic change in world-view would be necessary because it could very well signal the need for statistical laws denying knowledge of the cause for phenomena—a priori probabilities. For example, Ernest Rutherford's law for how many of a large number of atoms will undergo radioactive decay in a certain time period is a statistical law, but not a priori probabilistic. Poincaré emphasized that the laws governing the behavior of *individual* atoms would turn out to be causal laws, and this faith would be realized by future analyses of Rutherford's recently proposed model of the atom as a miniature Copernican system.[13]

In contrast to the continuous spectrum of cavity radiation stood the discrete spectrum of free atoms. For if a gas of, for example, hydrogen is heated it emits light whose spectrum is a series of unequally spaced lines differing in brightness and frequency. The spectral lines of hydrogen were especially well known, and by the end of the first decade of the twentieth century there were formulae without theoretical foundation that served to classify them. In 1913, Bohr offered a theory from which these series could be deduced.[14]

Bohr's theory was predicated explicitly upon discontinuities in nature and in its laws, as well as violating both classical mechanics and electrodynamics. The theory is based upon two postulates: First, in order to account for the stability of matter and to set a scale for the size of the atom, Bohr asserts that contrary to classical mechanics not every orbit is permissible for the planetary electron. Bohr refers to the permitted orbits as "stationary states"[15] which can be calculated from classical mechanics suitably altered to include Planck's constant h from his theory of the continuous spectrum of cavity radiation. In particular, there is a ground state below which the electron cannot descend. While in a stationary state an electron is in rotation about the nucleus, but it does not radiate as it should according to classical electrodynamics. Metaphorically Bohr describes the stationary states as "waiting places"[16] where the electron resides before making a transition to another stationary state. Neither classical mechanics nor electrodynamics can discuss the transition process. So Bohr proposed a second postulate: The transition downward of the electron is accompanied by the emission of radiation of energy $h\nu$, where ν is the frequency of the emitted radiation. For Bohr the quantum jump is an "essential discontinuity"[17] because it is demanded by the two postulates of the theory. Moreover, the quantum jump is unvisualizable. Consider, for example, the simplest element hydrogen which the Bohr theory considers as a solar system with one planet—the electron. The electron makes a quantum jump by disappearing from one stationary state and reappearing in another one, somewhat like the smile of the Cheshire cat.

Despite Bohr's bold departures from classical physics a picture survived of the atom as a miniature Copernican system, so familiar and so easy to visualize because, as Rudolf Arnheim observed of corpuscular models, there is a sharp difference between "figure" and "ground."[18] Bohr's theory achieved astonishing successes. For example, it could discuss quantitatively many of the chemical and physical characteristics of hydrogen as well as offer a means to relate the atomic structure and chemical properties of the more complex elements. By 1920, it had also disposed of the classical theory of the scattering of light by atoms, that is, dispersion. The classical theory of dispersion uses a model of the atom analogous to that of Planck: an atom reacting to light is represented by the easily visualizable model of charged oscillators. The charged oscillators are set into motion by the incident radiation thereby becoming sources of coherent secondary waves; and the frequencies of the spectral lines are closely approximated by the frequencies of the oscillators. On the other hand, according to Bohr's atomic theory, atoms are not charged oscillators but miniature planetary systems, and the frequency of a spectral line is not the frequency of an electron in a stationary state. By 1920, the matter had gone beyond the competition between two theoretical viewpoints because certain patterns of spectral lines from dispersion experiments could be accounted for only by the quantum theory (as Bohr's theory of the atom was referred to by then).[19] However, in 1920 the quantum theory could offer no detailed description of dispersion; this would have to await further development of an idea proposed in detail by Bohr two years earlier.

In 1918, Bohr extended the quantum theory by proposing a method to discuss the brightness, that is, intensity, of spectral lines.[17] It is based upon a limiting process which would be a guiding theme in the years ahead, and which he named in 1920 the "correspondence principle."[20] The correspondence principle is based upon the following observation. The spectral lines of hydrogen are unequally spaced, becoming ever closer with the lines' decreasing frequencies, until, to the naked eye, the spectrum becomes a blur of white. The white blur is the case expected from the viewpoint of classical electrodynamics according to which the planetary electron radiates continually and eventually spirals into the nucleus, but matter is stable and hence Bohr's first postulate. The low frequency spectral lines are the result of transitions downward between stationary states far from the nucleus. The correspondence principle asserts that in this region the classical and quantum predictions for the characteristics of the emitted radiation approach each other asymptotically. Thus, in this region to a high degree of accuracy the well known results of classical electrodynamics, suitably altered to contain quantum instead of classical frequencies, could be equated to the expressions offered by the quantum theory. Then, these equalities are assumed to hold reasonably well in the regions close to the nucleus. Bohr's problem was to find a quantum-theoretical description of the unvisualizable quantum jump. He found it in the *A* and *B* coefficients of Einstein's quantum theory of 1916–17 for the case of atoms in equilibrium with cavity radiation.[21] The *A* and *B* coefficients are the probabilities for an atom to make transitions that are spontaneous or induced by external radiation, respectively. Although Einstein's researches in the years 1905–16 are too often looked upon as revolutionary, he viewed them as extending classical physics. So it is not surprising that like Poincaré, Einstein considered the appearance of probabilities as a "weakness"[22] of his quantum theory of radiation. Bohr, on the other hand, took them over into his quantum theory of the atom as a priori probabilities in order to provide a mathematical description for the unvisualizable quantum jump. This was a courageous step for it widened the gap between the quantum theory and classical physics where there is no room for a priori probabilities.

In a major monograph of 1923, Bohr presented a scientific and epistemological analysis of the problems confronting the "possibility of forming a consistent picture [*Bild*] of phenomena" with which the principles of the quantum theory "can be brought into conformity."[23] Although Bohr's reason for writing it almost certainly concerned the withering away in 1921–23 of the picture of the atom as a miniature Copernican system, he does not consider of fundamental importance problems concerning the constitution of complex atoms. He emphasizes that despite the "good service" offered by the classical theory of radiation through the correspondence principle, "we must always keep clearly before us the far-reaching character of the departure from our customary ideas [*Vorstellungen vor Augen*] which is effected by the introduction of discontinuities." Among the immediate departures were the two postulates of Bohr's theory of 1913 and introduction of a priori probability to discuss the unvisualizable quantum jump in 1918;

in 1923, others would follow. In facing these problems Bohr warns the reader of the difficulties of extending our customary intuition into the microscopic domain: "we have in mind the fundamental difficulties which stand in the way of the effort to reconcile the appearance of discontinuities in atomic processes with conceptions of classical electrodynamics."

A method Bohr proposes for removing the "fundamental difficulties" concerns strictly maintaining the conservation laws of energy and momentum in "individual processes" using light quanta, that is, in the scattering of light by individual atoms. However, Bohr continues, this method cannot be "considered as a satisfactory solution" because the "picture [*Bild*]" of light quanta excludes the possibility of discussing interference. The success of the hypothesis of light quanta in accounting for certain experimental results led him to conclude that no matter which hypothesis on the nature of light would prove ultimately to be satisfactory, a space and time description of atomic processes could not "be carried through in a manner free from contradiction by the use of conceptions borrowed from classical electrodynamics." Since classical physics links the conservation laws of energy and momentum with the picture of particles moving continuously in space and time, then "we must be prepared for the fact that deduction from these laws will not possess unlimited validity." The correspondence principle rather than the conservation laws would be a guide to a consistent quantum theory.

Bohr's other method for removing fundamental difficulties concerns a new approach to the problem of dispersion that will dramatically alter the course of quantum theory. Toward using the correspondence principle to extend the classical theory of dispersion into the quantum theory, Bohr writes that the atom responds to external radiation like a "number of classical oscillators" whose frequencies are those of the observed spectral lines. He refers to this response as the "coupling mechanism," and cites recent work by Rudolf Ladenburg who independently had a similar "train of thought" in 1921.[24] Then, in 1923, Ladenburg applied the correspondence principle to the classical theory of dispersion in order to formulate a mathematical version of the coupling mechanism.[25] In summary, the new theory of dispersion was not based upon the picture of the atom as a miniature Copernican system.

By 1924, Arthur Compton's interpretation for the change of wavelength of very high frequency radiation (X rays) scattered from metal foils furnished further evidence for the reality of the light quantum.[26] Among the physicists who disagreed with Compton was Bohr for whom the tension between the two conceptions of light would have to be resolved on the basis of a wave theory. For although there may be essential discontinuities in individual processes, our customary intuition demands that light be a continuous phenomenon.

Bohr's viewpoint appeared in a paper of 1924 coauthored with H. A. Kramers and John C. Slater, but almost certainly written entirely by Bohr.[27] They consider this paper to be a "supplement" to Bohr's monograph of 1923, and the solution to the "fundamental difficulties" which stand in the way of reconciling classical electrodynamics with the essential discontinuities of atomic physics. Their solution is based upon Bohr's coupling mechanism which asserts that the atom reacts to incident radiation like a "set of 'virtual oscillators' " with the proper quantum frequencies, and "such a picture [*Bild*] was used by Ladenburg." Bohr's use here of the notion of a "picture" is in the sense of the interpretation offered by Ladenburg's mathematical framework, and this differs from the notion of "picture" in the essay of 1923 where it was meant as visualization. Indeed, one cannot visualize the planetary electron in a stationary state in the hydrogen atom represented by as many oscillators as there are transitions from this state. The set of virtual oscillators replacing the image of the planetary electron in a stationary state continually emits a virtual radiation field transporting only the probability for an electron to make a transition; a priori probability is introduced into the theory through the correspondence principle which contains the A and B coefficients. Thus, for example, since the virtual radiation field of one atom can induce a transition upward in another atom without undergoing the corresponding downward transition, Bohr, Kramers, and Slater abandon "any attempt at a causal connexion between transitions in distant atoms," as well as energy and momentum conservation in the individual interactions "so characteristic of the classical theories." Nevertheless, they continue, energy and momentum are conserved statistically, that is, for many scattering processes (this contrasts with Poincaré's assessment of laws like that of radioactive decay).

In contrast to Compton's "formal interpretation" using light quanta, and hence energy and momentum conservation, Bohr, Kramers, and Slater consider the Compton effect as a continuous process. Each illuminated electron in the target emits coherent secondary wavelets which can be understood as light scattered from a virtual oscillator. This description displaced physical reality further from the "classical conceptions" because the scattered electron and virtual oscillator need not have the same velocity or position. Additionally, since the virtual radiation field transports probability, they predicted that the scattered electron possesses the probability of having momenta in any direction.

Thus, in order to maintain the wave concept of radiation Bohr was willing to pay a high price—namely, relinquishing the picture of an electron as a localized quantity, the laws of energy and momentum conservation and causality. This was a desperate time for Bohr because he could very well have believed that the hypothesis of the virtual oscillators was the last gasp of his program for a description of the interaction of light with matter that was macroscopically continuous. He as much as admits that this is a physics of desperation in the Bohr, Kramers, and Slater paper of 1924, where he writes that "it seems at the present state of science hardly justifiable to reject a formal interpretation as that under consideration as inadequate." The violation of energy and momentum conservation was treated with caution by most physicists; Born, Heisenberg, and Kramers omitted it from their subsequent research based upon the virtual oscillators.[28] The experimentalist E. N. da C. Andrade caught the mood of the times in the third edition of his book *The Structure of the Atom* which appeared in late 1926. There he considered the suggestion of nonconservation of energy as "repugnant."[29]

Although it is not my goal here to discuss the nature of Heisenberg's discovery of the new quantum mechanics in mid-1925, nevertheless a few words are necessary on the thinking that led him to it. The reason is that to Heisenberg during the period mid-1924 to mid-1925 the virtual oscillators went from being merely appealing to a guiding theme.[30] But we must keep in mind that the hypothesis of virtual oscillators was treated purely formally in the paper of Bohr, Kramers, and Slater, and then in subsequent elaborations using the Bohr, Kramers, and Slater theory of Ladenburg's preliminary theory of dispersion by Born and Kramers. Furthermore, Born and Kramers had been unable to extract from their formalisms assertions concerning empirical data.

My research has led me to a missing piece in the jigsaw of Heisenberg's discovery—his paper of 1925, "On an Application of the Correspondence Principle to Problems Concerning the Polarization of Fluorescent Light"[31] and its analysis of the dispersion experiment of the eminent American physicist R. W. Wood with A. Ellett,[32] which in retrospect must be considered as a turning point in the genesis of the quantum theory. In the paper of 1925, Heisenberg used the virtual oscillators to free himself from the notion of electrons in stationary states that are planetary orbits, for the "virtual oscillators are connected only in a very symbolic manner" with electrons in such states. Of great importance to Heisenberg was that he could use the virtual oscillators to make certain predictions that agreed adequately with the data of Wood and Ellett. He recollected that this paper showed the "necessity for detachment from intuitive models."[33] It was clear that the set of virtual oscillators were "more real than the orbit";[34] indeed, although particle tracks had been observed in cloud chambers, an electron in a planetary orbit had never been observed. The virtual oscillator representation would be Heisenberg's starting point in the seminal paper on the new quantum mechanics.[35] There the unvisualizable electron is characterized by relations among observable quantities, for example, the frequency of the atom's spectral lines. When Pauli, early in 1926, deduced the Balmer formula from Heisenberg's theory (that is, one of the spectral line series for the hydrogen atom), he emphasized at the outset in words conflicting with his reminiscence of 1955 (see Introduction) that "Heisenberg's form of the quantum theory completely avoids a mechanical-kinematical demonstration [*Veranschaulichung*] of the notion of electrons in the stationary states of the atom."[36]

Although the renunciation of a picture of the bound electron was a necessary prerequisite to the discovery of the new quantum mechanics, nevertheless the lack of *Veranschaulichung* or *Anschaulichkeit* or an intuitive [*anschauliche*] interpretation was of great concern to Bohr, Born, and Heisenberg, and this concern emerges from their scientific papers of the period 1925–27. For example, in the important paper of Born, Jordan, and Heisenberg of late 1925, Heisenberg writes in the "Introduction" that the present theory labors "under the disadvantage of not being directly amenable to a geometrically intuitive interpretation [*anschauliche interpretiert*] since the motion of electrons cannot be described in terms of the familiar concepts of space and time."[37] Heisenberg continues, "In the further development of the theory, an important task will lie in the closer investigation of the nature of this correspondence between classical and quantum mechanics and in the manner in which symbolic quantum geometry goes over into intuitive classical geometry [*anschaulich klassiche Geometrie*]."[38] In the section, "The Zeeman effect", probably also written by Heisenberg, the notion of planetary stationary states arises when he writes of the inability of the new quantum mechanics to resolve the problem of the anomalous Zeeman effect as perhaps due to the result of an "intimate connection between the innermost and outermost orbits . . ."; however, Heisenberg hoped that the recent hypothesis of an electron spin by Uhlenbeck and Goudsmit might provide an alternate route.[39] Yet this fourth degree of freedom for the electron could not be visualized because a point on a spinning electron of finite extent could move faster than light.

Born also felt uncomfortable without a means to visualize the concepts in the mathematical formalism of the new quantum mechanics. In late 1925, Born writes that "we have the right to use the terms 'orbit' or even 'ellipse,' 'hyperbola,' etc. in the new theory. . . . Our imagination is restricted to a limiting case of possible physical processes."[40]

Andrade, in his book published in late 1926, tempered his praise of the new quantum mechanics by pointing out that "its weakness is that we have no geometrical or mechanical picture" of the physical processes to which it is applied of the type to which "we are accustomed."[41]

In the latter part of 1925, Bohr acknowledged the empirical disproof of the prediction of Bohr, Kramers, and Slater for the nonconservation of energy and momentum in the Compton effect.[42] The experimental data[43] showed that if the initial energy and momentum of the light quantum is known, then the final values of these quantities for the scattered light quantum and electron could be predicted (initially the target electron is at rest). Nevertheless, Bohr steadfastly maintained his traditionally dualistic viewpoint by asserting that the data did not decide definitely between "two well-defined conceptions of light propagation in empty space." Rather the identification of a "coupling of individual processes . . . forced upon the picture a corpuscular transmission of light." In Bohr's view accompanying the inclusion of light quanta are "unavoidable fluctuations in time" for systems in small volumes, making it even more difficult to use "intuitive pictures [*anschaulicher Bilder*]"[44] to discuss collision processes and the structure of atoms. Of importance to Bohr was recent theoretical work offering further evidence for the renunciation of pictures in space and time—the researches of L. de Broglie of 1924 and of Einstein in 1924–25.[45] De Broglie proposed a wave-particle duality for matter. A result of Einstein's theory of the ideal quantum gas was the indistinguishability of its particles. This was a dramatic event for, in addition to the bound electron being unvisualizable, free particles have now lost their individuality.

Acknowledging Heisenberg's new quantum mechanics, Bohr writes that it may at first "seem deplorable" that in atomic physics "we have apparently met with such a limitation in our usual means of visualization"; nevertheless, "mathematics in this field too, presents us with tools to prepare the way for further progress."[46] Thus, where visualization has been lost, mathematics must be the guide toward "further progress." In summary, by the latter part of 1925, empirical data and theoretical results caused Bohr to demarcate between the conservation laws and pictures in space and time.

Visualization Regained (1926–27)

Erwin Schrödinger, in the third of the four "communications" of 1926, left no room for doubt that a sense of aesthetics inspired him to formulate the wave mechanics;[47]

My theory was inspired by L. de Broglie, *Ann. de Physique* (10) 3, p. 22, 1925 (Thèses, Paris, 1924) and by short but incomplete remarks by A. Einstein, *Berl. Ber.* (1925) pp. 9 ff. No genetic relationship whatever with Heisenberg is known to me. I knew of his theory, of course, but felt discouraged, not to say repelled, by the methods of transcendental algebra, which appeared very difficult to me and by the lack of visualisability [*Anschaulichkeit*].

In a more objective tone one of his principal criticisms against the quantum mechanics is that it appeared to him "extraordinarily difficult" to approach such processes as collision phenomena from the viewpoint of a "theory of knowledge" in which we "suppress intuition [*Anschauung*] and operate only with abstract concepts such as transition probabilities, energy levels, and the like." For although, he continues, there may exist "things" which cannot be comprehended by our "forms of thought," and hence do not have a space and time description, "from the philosophic point of view" Schrödinger was sure that "the structure of the atom" does not belong to this set of things.

Another of Schrödinger's reasons for preferring a wave-theoretical approach is his preference for a continuum-based theory in which he claimed there are no quantum jumps, over the "true discontinuum theory" of Heisenberg. In addition, Schrödinger pushed his proof of the mathematical equivalence of the wave and quantum mechanics to the conclusion natural to his viewpoint—when discussing atomic theories he "could properly also use the singular."

But what sort of picture did Schrödinger offer? He maintains that no picture at all is preferable to the miniature Copernican atom, and in this sense the purely positivistic standpoint of the quantum mechanics is preferable because of "its complete lack of visualization"; however, this conflicts with Schrödinger's philosophic viewpoint. Schrödinger bases his visual representation of bound and free electrons on the comparison with classical electrodynamics of the solution to the fundamental wave equation of the wave mechanics, that is, the wave function. The electron in a hydrogen atom is represented as a distribution of electricity around the nucleus. However, Schrödinger's proof of the localization of a free electron represented as a packet of waves[48] was shown by Heisenberg to be invalid;[49] rather, the packet of waves does not remain localized. In addition, Schrödinger does emphasize that his visual representation is unsuitable for systems containing two or more electrons because the wave function must be represented in a space of 3N-dimensions, where N is the number of particles.

To summarize the state of quantum physics in the first half of

1926: While no adequate atomic theory existed as of July 1925, by mid-1926, there were two seemingly dissimilar theories. Quantum mechanics was purported to be a "true discontinuum theory."[50] Although a corpuscular-based theory, it renounced any visualization of the bound corpuscle itself (at this time quantum mechanics could discuss only bound state problems). However, its mathematical apparatus was unfamiliar to physicists and also difficult to apply. On the other hand, there was Schrödinger's wave mechanics which was a continuum theory focusing entirely upon matter as waves, offering a visual representation of atomic phenomena and accounting for discrete spectral lines. Its more familiar mathematical apparatus set the stage for a calculational breakthrough. The wave mechanics delighted the more continuum-based portion of the physics community, particularly Einstein and Planck.[51] Although the final experimental verification of the complete wave-particle duality of matter would not appear until 1927, many already subscribed to it.

To the best of my knowledge Heisenberg's first published response to Schrödinger's wave mechanics is the paper of 1926, "Many-Body Problem and Resonance in Quantum Mechanics."[52] There Heisenberg explains that although the physical interpretations of the two theories differ, their mathematical equivalence allows this difference to be put aside; for "expediency" in calculations he will utilize the Schrödinger wave functions, with the caveat that one must not impose upon the quantum theory Schrödinger's "intuitive pictures [*anschaulichen Bilder*]." In fact, Heisenberg points out that such an attitude would prevent treating the many-body problem. To make this clear he refers to Schrödinger's viewpoint concerning the principal difficulty confronting either the quantum or wave mechanics in treating many-body systems; namely both theories use formulae from classical mechanics which discuss these interactions as if the atomic objects were point particles, but this "is no longer permissible" because point charges are "actually extended states of vibration which penetrate into one another."[53] Heisenberg's rebuttal in his paper *treating* the many-body problem is that we must set limitations "upon the discussion of the intuition-problem [*Anschauungsfrage*]," for there are cases in which the "wave representation is more constrained"—for example, "the notion of the spinning electron"[54] which resists perception or intuition through pictures. Heisenberg also objects that Schrödinger's method is not a consistent wave theory of matter in the sense of de Broglie whose waves are in a visualizable space of three dimensions.

The tension between the quantum and wave mechanics increased with the appearance of Born's quantum theory of scattering in the latter part of 1926.[55] Born's analysis of data from the scattering of electrons from hydrogen atoms convinced him of the need for a quantum-theoretical description of scattering consistent with the conservation of energy and momentum; yet, Born writes, neither scattering problems nor transitions in atoms can be "understood by the quantum mechanics in its present form"[56]—here, under quantum mechanics, Born includes both Heisenberg's and Schrödinger's mechanics. His reasons are: Heisenberg's quantum mechanics denies an "exact representation of the processes in space and time";[57] Schrödinger's wave mechanics denies visualization of phenomena with more than one particle. In Born's view treating problems concerning scattering and transitions requires the "construction of new concepts," and his vehicle would be Schrödinger's version because it allows the use of the "conventional ideas of space and time in which events take place in a completely normal manner," that is, the possibility of visualization.

One new concept that Born proposes is rooted in some unpublished speculations of Einstein; namely, that light quanta are guided by a wave field that carries only probability, providing a means to discuss interference and diffraction using light quanta. Born boldly assumes the "complete analogy" between a light quantum and an electron in order to postulate the interpretation that the "de Broglie-Schrödinger waves," that is, the wave function in a three dimensional space, is the "guiding field" for the electron. From this wave function can be found the probability for the electron to be in a certain region of space. Thus, Born's quantum theory of scattering combines the de Broglie-Schrödinger waves with corpuscles. This view led Born "to be prepared to give up determinism in the world of atoms."[58] For although the carrier of probability develops causally, that is, according to Schrödinger's equation, all final states are probable (although in general not equally probable) that are consistent with the conservation laws of energy and momentum. So, in Born's view, quantum mechanics distinguishes between causality and determinism (compare this with Poincaré's statement of indeterminism in the previous section).

But Born had to go even further, for since the quantum "jump itself" in an atom "defy all attempts to visualize it," then for transition processes, as they occur in atoms, the notion of causality is meaningless, and one is left only with the "quantum mechanically determinate."[59]

Heisenberg was enraged over Born's use of the Schrödinger theory and his assessment of the quantum mechanics. In fact, from the beginning Heisenberg had been no less outspoken than Schrödinger, for soon after the appearance of wave mechanics he referred to it as "disgusting" in a letter to Pauli.[60] Heisenberg recalled that "Schrödinger tried to push us back into a language in which we had to describe nature by '*anschauliche Methoden*'. . . . Therefore I was so upset about the Schrödinger development in spite of its enormous successes."[61] Then came Born's paper in which "he went over to the Schrödinger theory."

Heisenberg described these developments as very disturbing to his "actual psychological situation at that time"; namely, that quantum mechanics was a complete and thus closed system. Born, on the other hand, had assessed it as incomplete and introduced a new hypothesis using Schrödinger's wave mechanics. It was in response to this highly charged emotional atmosphere that Heisenberg wrote his paper of 1926 "Fluctuation Phenomena and Quantum Mechanics."[62] He remembered that although this paper received very little attention, "For myself it was a very important paper."[63] Indeed, it is a paper written by an angry man in which Born's theory of scattering is not cited and Schrödinger is sharply criticized. There Heisenberg demonstrated that a probability interpretation emerges naturally from the quantum mechanics and can be understood only if there are quantum jumps. At the conclusion he comes down firmly in favor only of a corpuscular viewpoint. This becomes crystal clear in Heisenberg's important review paper of 1926 "Quantum Mechanics"[64] where once again Born is not mentioned and Schrödinger is soundly criticized.

"Our customary intuition [*Anschauung*]," Heisenberg begins, is contradicted by the phenomena occurring in small volumes where there is the "typically discontinuous element," and where a concept such as that of the light quantum has proven fertile. He continues with a statement carrying an implicit rebuke of Born's theory of scattering: "Nevertheless, in contrast to material particles, we have never before attributed the kind of reality to light quanta which befits objects of the everyday world." His reason is that the light quantum contradicts the "known laws of optics," for example, interference phenomena. On the other hand, Heisenberg continues, perhaps it is as Einstein has emphasized that "conversely the *electron* is due a similar degree of reality as the light quantum" (italics in original). Heisenberg considers this point as symptomatic of the fundamental problem of atomic physics—namely, "the investigation of that typically discontinuous element and of that 'kind of reality.' "

Heisenberg's "*first decisive restriction* in the discussion of the reality of corpuscles" (italics in original) is the renunciation of the notion of the electron's position in an atom because this was necessary in order to formulate the quantum mechanics in terms of relationships among observable quantities. This procedure has its dangers, Heisenberg continues, for the new theory renounces "visualization [*Anschaulichkeit*]" thus running the risk of having internal contradictions.

Heisenberg cites as a "further restriction" on the reality of the corpuscle a result from Einstein's theory of the ideal quantum gas—"the individuality of a corpuscle is lost."

Whereas Heisenberg began this essay by demonstrating how nature in the small contradicts our customary intuition he concludes with emphasis upon the fact that the existing scheme of quantum mechanics contains contradictions of the "intuitive interpretations [*anschaulichen Deutungen*]" of different phenomena, and this is not satisfactory. For the quantum theory places restrictions on the reality of the corpuscle and there lurks the notion of the light corpuscle. Heisenberg's essay ends with a most interesting and curious passage. Despite repeated warnings throughout this paper and in the many-body paper against intuitive interpretations of the quantum mechanics, Heisenberg reports from Copenhagen where he and Bohr are in the midst of their intense struggle toward a physical interpretation of the quantum mechanics (late fall 1926–spring 1927): "Hitherto there is missing in our picture [*Bild*] of the structure of matter any substantial progress toward a contradiction-free intuitive [*anschaulichen*] interpretation of experiments which in themselves are contradiction-free."

It was on this note that Heisenberg began his publication of 1927 on the fundamental problems of the quantum mechanics, "On the Intuitive [*anschaulichen*] Content of the Quantum-Theoretical Kinematics and Mechanics,"[49] where he brings to a conclusion the analysis begun in "Quantum Mechanics." But what kind of "intuitive content" can a physicist offer who denies visualization of physical processes? In the very first sentence Heisenberg begins his bold reply to this question by stating two criteria for a theory "to be understood intuitively": in all simple cases the theory's experimental consequences can be thought of in a qualitative manner; and its application should lead to no internal contradictions. Although the mathematical scheme of quantum mechanics requires no revisions, Heisenberg continues, "Heretofore, the intuitive [*anschauliche*] interpretation of the quantum mechanics is full of internal contradictions which become apparent in the struggle of the opinions concerning discontinuum- and continuum-theory, waves and corpuscles." Thus, the quantum mechanics satisfies only one of Heisenberg's criteria "to be understood intuitively." Heisenberg writes that just as in the general relativity theory, where the extension of our usual conceptions of space and time to very large volumes follows from the mathematics of the theory, a revision of our usual kinematical and mechanical concepts "appears to follow directly from the fundamental equations of the quantum mechanics." Heisenberg deduced from these equations the result that, unlike in classical physics, in the atomic domain the uncertainties in measuring the position and momentum of an atomic particle cannot be simultaneously reduced (even in principle) to zero. Rather, the product of the uncertainties is a small but nonzero number—Planck's constant. For example, the more precisely the particle's position is measured, the less precisely can its momentum be ascertained. Heisenberg attributed the cause of this uncertainty relation to be the "typical discontinuities" which contradict our customary intuition. He concludes that the mathematical formalism of quantum mechanics determines the restrictions in the atomic domain of such classical concepts as position and momentum. Clearly, in order to maintain his aesthetic of a limited wave-particle duality which is not linked to visualization, Heisenberg has boldly demarcated the notion of "to be understood intuitively" from the visualization of atomic processes. He has chosen to resolve the "struggle of the opinions" between thema-antithema couples with a corpuscular-based theory that lacks visualization of the corpuscle and severely restricts visualization of physical processes.

This viewpoint led him to assert that in the case of "the strong formulation of the causal law: 'if we know the present exactly, then we can calculate the future,' it is not the consequent that is false but the presupposition." The "strong formulation of the causal law" is the one from classical physics, and it is dependent upon visualization and the absence of discontinuities. However, the uncertainty relation between position and momentum places limits upon the precision with which the initial conditions of an atomic particle can be specified; its path cannot be traced to any arbitrary degree of accuracy as in classical physics. Heisenberg's rejection of the causal law from classical physics is therefore not unexpected.

With these results, Heisenberg concludes with a remarkable passage, "one will not have to any longer regard the quantum mechanics as unintuitive [*unanschaulich*] and abstract."

Heisenberg's strong predisposition to his successful quantum mechanics predicated upon a lack of visualizability is a reflection of his preference for nonvisual thinking, and is undoubtedly the root of his redefinition of intuition. His lack of trust in visual thinking to understand quantum mechanics could very well have been further reinforced by discussions with Bohr during the period of their intense struggle to understand the riddles of the quantum theory. For imagine using visual representations, as Heisenberg writes in the uncertainty principle paper, to think qualitatively of the experimental results in all simple cases—for example, the determination of the slit through which an electron passes in the diffraction of a low intensity beam of electrons by a double slit grating, or what it means for a light quantum to be polarized.[65] Heisenberg recalled that "we couldn't doubt that this [i.e., quantum mechanics] was the correct scheme but even then we didn't know how to talk about it"; these discussions left them in "a state of almost complete despair."[66] The loss of visualization must have been especially difficult for Bohr whose essays are filled with visual words—for example, picture [*Bild*], visual ideas [*Vorstellungen vor Augen*], mechanical models. Arnheim, in an essay on the psychology of art, writes of the "apprehension" that develops in a scientist during a transition from a corpuscular theory with "determined contour line" to more complex models.[67] The states of mind of both Bohr and Heisenberg fit this description, for they were set adrift, Bohr lacking visualization and both Bohr and Heisenberg distrusting their intuitions.

Heisenberg recalled that their approaches to gedanken experiments differed. Bohr, by late 1926, had accepted the duality in the quantum theory and its reflection in nature as the complete wave-particle duality, even though the wave aspect of matter had not yet been definitely established experimentally. For Bohr the wave-particle duality was the "central point in the whole story,"[68] because it permitted him to use visual thinking once again, that is, to play with pictures of waves and particles.

Heisenberg relied solely upon the mathematical scheme of quantum mechanics until December of 1926 when he became aware of the Dirac and Jordan transformation theories, proving to his satisfaction the equivalence between the wave and quantum mechanics.[69] Then Heisenberg could understand Bohr's interchangeable use of wave and quantum mechanics because he could mathematize Bohr's visual arguments. Nevertheless, Heisenberg steadfastly refused to believe that there was a complete dualism in quantum theory, rather the transformation theory showed how "very flexible"[70] was the quantum mechanics.

In the light of their different viewpoints, Bohr's dissatisfaction with Heisenberg's uncertainty principle paper should come as no surprise. Bohr insisted that one cannot allow the mathematical formalism to restrict words like position and momentum because, despite the uncertainty relation, you have to use them "just because you haven't got anything else."[71] An unpleasant atmosphere developed in which Heisenberg refused to change the content of the paper but did acquiesce to add a "*Nachtrag bei Korrectur*" (Postscript with Corrections).[72] There Heisenberg writes that

Bohr's recent investigations led him to conclude that uncertainty in observation is not rooted "exclusively upon the presence of discontinuities," but in the wave-particle duality of matter.

On 16 September 1927 at the International Congress of Physics at Como, Italy, Bohr presented a viewpoint which he hoped would "be helpful" to "harmonize the apparently conflicting views taken by different scientists."[73] He realized that the "classical mode of description must be generalized" because our customary intuition cannot be extended into the atomic domain. There Planck's constant links the measuring apparatus to the physical system under investigation in a way that is "completely foreign to the classical theories," and this is the root of the unavoidable statistics of the quantum theory. While in the classical theories this interaction can be neglected, in the atomic domain it cannot. Then, in the atomic domain, the notion of an undisturbed system developing in space and time is an abstraction and "there can be no question of causality in the ordinary sense of the word," that is, strong causality. Bohr's viewpoint on this situation is the complementarity principle:

> The very nature of the quantum theory thus forces us to regard the space-time coordination and the claim of causality, the union of which characterizes the classical physical theories, as complementary but exclusive features of the description, symbolizing the idealization of observation and definition respectively.

Bohr's response is to separate the causal law from a space-time description; the union of the two was the classical or strong causality. The causal law and space-time pictures are complementary because they are both necessary for a complete description of phenomena.

Bohr next emphasizes that his deliberations upon the relationship between visual thinking and the wave-particle duality of light were central to the genesis of the complementarity viewpoint. Light as a wave is useful for describing interference phenomena. But the conservation of energy and momentum in the individual interactions of light with atoms, for example, the Compton effect, "finds its adequate expression" in the light quantum (recall that in classical physics also these conservation laws were linked to particles). Since a description of the characteristics of a light quantum involves Planck's constant, tracing its path means an interaction with a measuring apparatus and so we are "confined to statistical considerations." Bohr continues: "This situation would seem clearly to indicate the impossibility of a space-time description of light phenomena." This was Bohr's view of light quanta in 1923 and in the face of experimental data in 1925 as well. Here, however, Bohr associates causality with the predictive powers of the conservation laws of energy and momentum, rather than with space-time pictures that illustrate predictions for the position of the light quantum. Since the wave-particle duality holds also for atomic entities, for example, the electron, then space-time pictures cannot be associated with a causal description of the interactions among these particles. In the atomic domain the quantities that characterize a particle (energy and momentum) are connected through Planck's constant with the quantities that characterize a wave (frequency and wavelength). However, in the physics of macroscopic bodies this connection has been neglected because of the insensitivity of our perceptions to the effects of a physical constant as small as Planck's constant; for this reason the wave and particle pictures seem contradictory. According to the viewpoint of complementarity, the pictures of light and matter as waves and particles are not contradictory, as had been thought previously, but are "complementary pictures" because they are both necessary for a complete description of atomic phenomena; they are mutually exclusive because in a given experimental arrangement atomic entities can exhibit only one of their two sides. The scheme of complementarity permits a self-consistent description of how the quantum theory relates to the simple experiments that had driven Bohr and Heisenberg to despair. For in the complementarity paper Bohr emphasizes that "every word in the language refers to our ordinary perception," and according to our ordinary perception there are only two kinds of phenomena—corpuscular and undulatory—just as our intuition tells us that "things" are either discontinuous or continuous. The failure of our ordinary intuition in the atomic domain, Bohr writes, is rooted in "the general difficulty in the formation of human ideas, inherent in the distinction between subject and object."

This completes Bohr's analysis of 1927, which contains the viewpoint soon to be referred to as the *Kopenhagener Geist der Quantentheorie.*[74] It is an extraordinary analysis because Bohr's method to arrive at a contradiction-free interpretation of the quantum theory led him to ever-deeper levels of analysis: from a purely scientific analysis, to an epistemological analysis, to an analysis of perceptions, and then to the origins of scientific concepts. A necessary prerequisite to this analysis was the acceptance of the complete wave-particle duality in nature. This permitted Bohr to use the symmetry of the pictures of waves and particles which are familiar from our customary intuition. He could then discuss all simple experiments. Visual thinking preceded verbal thinking, and linked with visual thinking was Bohr's new aesthetic of the symmetry of pictures afforded by accepting the complete wave-particle duality. Bohr's viewpoint stands in contrast to the views of Heisenberg and Born. Although Heisenberg's viewpoint maintained the validity of the conservation of energy and momentum, it was corpuscular-based and severely restricted visualization as well as the use of words such as position and momentum. Heisenberg considered the classical law of causality which is linked to the pictures to be invalid in the atomic domain. Born's viewpoint of 1926 contained energy and momentum conservation, visualization, waves and particles, but the waves were not observable quantities. Furthermore, defining causality according to Schrödinger's equation as the development in space and time of the probability waves led Born to accept the validity of causality in scattering processes and its lack of meaning in atomic transition processes. However, Bohr realized that only the choice of a complete dualism, vis-à-vis the holistic view of Heisenberg and the incomplete dualism of Born could offer the visualization associated with the customary intuition rejected by Heisenberg's holistic view. In addition, this choice led Bohr to a new causal law that is separated from visualization: space-time pictures of atomic processes can be constructed using concepts from our customary intuition, but one must understand that they are not causal descriptions. There is one point, however, shared by the views of Born and Bohr; namely, that contrary to the classical physics, the quantum theory is causal but not deterministic.

A study of Heisenberg's writings in the period 1928–29 in conjunction with his reminiscences reveals, not unexpectedly, that he understood the symmetry of the "complementary picture" from the mathematical statements of the complementarity principle—the methods of the quantization of wave fields formulated for the electromagnetic field by Dirac in 1927 and then extended to particles with mass by Jordan, Klein, and Wigner in 1927–28.[75] This is a form of the quantum theory in which the wave and particle aspects of matter can be transformed mathematically into one another and yet remain mutually exclusive. A fine illustration of the different modes of thinking of Heisenberg and Bohr is in Heisenberg's recollection of a conversation they had in the period just after the appearance of the papers of Jordan, Klein, and Wigner[76]:

> So the symmetry was complete only after these papers. Well Bohr perhaps in the first moment did not feel exactly that way, but I think later on he saw quite well that this was an illustration that he wanted.

In conclusion, the importance for creative thinking of the domain where art and science merge has been emphasized by the great philosopher-scientists of the twentieth century—Bohr, Einstein, and Poincaré. For in their research the boundaries between disciplines are often dissolved and they proceed neither deductively through logic nor inductively through the exclusive use of empirical data, but by visual thinking and aesthetics.

Acknowledgments

This essay is based upon a longer study written during the summer of 1976 at Harvard University. I am deeply grateful to Professor Gerald Holton for the hospitality he has extended to me at Harvard and for his encouragement and support. I thank two of the pioneers of the quantum theory at Harvard, Professors E. C. Kemble and J. H. Van Vleck, for several enlightening conversations on physics, circa 1925.

Quotations from the *Archive for History of Quantum Physics* are taken from interviews of Werner Heisenberg by Thomas S. Kuhn and are on deposit at the *American Institute of Physics* in New York City, the *American Philosophical Society in Philadelphia*, the University of California in Berkeley, and at the *Niels Bohr Institute* in Copenhagen.

Notes and References

1. Niels Bohr, "The Atomic Theory and the Fundamental Principles underlying the Description of Nature," in Niels Bohr, *Atomic Theory and the Description of Nature* (1934; rpt. Cambridge: At the University Press, 1961), pp. 102–119. The quotation is on p. 108.

2. Wolfgang Pauli, "Exclusion Principle, Lorentz Group and Reflection of Space-Time and Charge," *Niels Bohr and the Development of Physics: Essays dedicated to Niels Bohr on the occasion of his seventieth birthday*, ed. Wolfgang Pauli (New York: Pergamon Press, 1955), pp. 30–51. The quotation is on p. 30.

3. *"Anschauung"* is something superior to merely viewing with the senses. It is a kind of sight abstracted from visualization of physical processes in the world of perceptions. For a discussion of the use of *"Anschauung"* in German philosophical writings in the nineteeth century see John Theodore Merz, *A History of European Thought in the Nineteenth Century* (1904–12; New York: Dover, 1965), vol. III, pp. 205, 445 and vol. IV, pp. 314–315.

4. Gerald Holton, *Thematic Origins of Scientific Thought: Kepler to Einstein* (Cambridge: Harvard University Press, 1973).

5. Mogens Anderson, "An Impression," *Niels Bohr: His Life and Work as Seen by His Friends and Colleagues*, ed. S. Rozental (New York: Interscience, 1967), pp. 321–324. The quotation is from p. 322.

6. For discussions of Planck's researches of 1900 see Max Jammer, *The Conceptual Development of Quantum Mechanics* (New York: McGraw-Hill, 1966), esp. ch. 1 and Martin J. Klein "Max Planck and the beginnings of the quantum theory," *Archive for History of Exact Sciences*, 1 (1962): 459–479.

7. Albert Einstein, "Über einen die Erzeugung und Verwandlung des Lichtes betreffenden heuristischen Gesichtspunkt," *Annalen der Physik*, 17 (1905): 132–148. For further discussions of Einstein's paper see Jammer, ch. 1; Holton, "On the Origins of the Special Relativity Theory," in ref. 4, pp. 165–183, esp. pp. 167–169; Martin J. Klein, "Einstein's First

Paper on Quanta," *The Natural Philosopher*, 2 (1963): 59–86; and Arthur I. Miller, "On Einstein, light quanta radiation and relativity in 1905," *American Journal of Physics*, 44 (1976): 912–923.

8. Max Planck, "Zur Theorie der Wärmestrahlung," *Annalen der Physik*, 31 (1910): 758–767.

9. Albert Einstein, "Die Plancksche Theorie der Strahlung und die Theorie der spezifischen Wärme," *Annalen der Physik*, 22 (1907): 180–190. For further discussion see Martin J. Klein, "Einstein, Specific Heats and the Early Quantum Theory," *Science*, 148 (1965): 173–180.

10. Henri Poincaré, "The Quantum Theory," in Henri Poincaré, *Mathematics and Science: Last Essays,* trans. John W. Bolduc (New York: Dover, 1963), pp. 75–88.

11. See Arthur I. Miller, "Poincaré and Einstein: A Comparative Study," to be published in *Boston Studies in the Philosophy of Science*, vol. XXXI.

12. Poincaré, *Last Essays*, p. 88.

13. Poincaré, "The Relations between Ether and Matter," *Last Essays*, pp. 89–101, esp. p. 93.

14. Niels Bohr, "On the Constitution of Atoms and Molecules," *Philosophical Magazine*, 26 (1913): 1–25, 476–502, 857–875; reprinted with an introduction by Léon Rosenfeld in *On the Constitution of Atoms and Molecules* (New York: Benjamin, 1963). For discussions of these papers see Jammer, esp. ch. 2 and John L. Heilbron and Thomas S. Kuhn, "The Genesis of the Bohr Atom," *Historical Studies in the Physical Sciences*, ed. Russell McCormmach (Philadelphia: University of Pennsylvania Press, 1969), vol. I, pp. 211–290.

15. Bohr, *Atoms and Molecules*, p. 7 of the reprint volume.

16. Niels Bohr, "On the Spectrum of Hydrogen," an Address delivered before the Physical Society of Copenhagen on 20 December 1913 and reprinted in Niels Bohr, *The Theory of Spectra and Atomic Constitution* (Cambridge: at the University Press, 1924), pp. 1–19. The quotation is from p. 11.

17. Niels Bohr, "On the Quantum Theory of Line-Spectra," Kgl. Danske Vid. Selsk. Skr. nat.-mat. Afd., series 8, IV (1918–1922): 1–118. This paper appeared in 1918 and is reprinted in *Sources of Quantum Mechanics*, ed. B. L. Van der Waerden (New York: Dover, 1967) pp. 95–136. The quotation is from p. 99 of the reprint volume.

18. Rudolf Arnheim, *Visual Thinking* (Berkeley: University of California Press, 1971), p. 286.

19. Niels Bohr, "On the Series Spectra of the Elements," an Address delivered before the Physical Society in Berlin 27 April 1920 and reprinted in Bohr, *Theory of Spectra*, pp. 20–60, esp. pp. 32–34.

20. Bohr, *Theory of Spectra*, p. 27.

21. Albert Einstein, "On the Quantum Theory of Radiation," translated in *Sources of Quantum Mechanics*, pp. 63–77. (Originally published in *Physikalische Zeitschrift*, 18 (1917): 121–128.) For further discussion of Einstein's paper see Jammer, ch. 3 and Martin J. Klein, "Einstein and the

Wave-Particle Duality," *The Natural Philosopher*, 3 (1964): 3–49.

22. Einstein, *Sources*, p. 76.

23. Niels Bohr, "On the Application of the Quantum Theory to Atomic Structure: Part I. The Fundamental Postulates of the Theory," *Proc. Cambr. Phil. Soc.* (Supplement) (1924). (Originally published in *Zeitschrift für Physik*, 13 (1923): 117–165.) Hereinafter all German words appearing in square brackets are from the German version of the work under discussion.

24. Rudolf Ladenburg, "The Quantum-Theoretical Interpretation of the Number of Dispersion Electrons," translated in *Sources of Quantum Mechanics*, pp. 139–157. (Originally published in *Zeitschrift für Physik*, 4 (1921): 451–468.)

25. Rudolf Ladenburg and Fritz Reiche, "Absorption, Zerstreuung und Dispersion in der Borschen Atomtheorie," *Die Naturwissenschaften*, 11 (1923): 584–598.

26. Arthur H. Compton, "A Quantum Theory of the Scattering of X-Rays by Light Elements," *Physical Review*, 21 (1923): 483–502. For a discussion of Compton's experiments see Roger H. Stuewer, *The Compton Effect: Turning Point in Physics* (New York: Science History Publications, 1975).

27. Niels Bohr, H. A. Kramers, and John C. Slater, "The Quantum Theory of Radiation," *Sources of Quantum Mechanics*, pp. 159–176. (Originally published in *Zeitschrift für Physik*), 24 (1924): 69–87 and *Philosophical Magazine*, 47 (1924): 785–802.) For further discussions of this paper see Jammer, ch. 4; Stuewer; and Klein, "The First Phase of the Bohr-Einstein Dialogue," *Historical Studies*, vol. II, pp. 1–39.

28. For example, H. A. Kramers, "The Law of Dispersion and Bohr's Theory of Spectra," *Sources of Quantum Mechanics*, pp. 177–180 (originally published in *Nature*, 113 (1924): 673–676), and Max Born, "On Quantum Mechanics," translated in *Sources*, pp. 181–198 (originally published in *Zeitschrift für Physik*, 26 (1924): 379–395).

29. E. N. da C. Andrade, *The Structure of the Atom* (New York: Harcourt, Brace and Company, 1926), p. 685.

30. See, for example, *Archive for History of Quantum Physics,* Interview with W. Heisenberg on 13 February 1963, p. 8.

31. Werner Heisenberg, "Über eine Anwendung des Korrespondenzprinzips auf die Frage nach der Polarisation des Fluoreszenzlichtes," *Zeitschrift für Physik*, 31 (1925): 617–626. In a forthcoming publication this paper will be discussed in detail.

32. R. W. Wood and A. Ellett, "On the Influence of Magnetic Fields on the Polarisation of Resonance Radiation," *Proc. Roy. Soc. London* (A), 103 (1923): 396–403 and R. W. Wood and A. Ellett, "Polarized Radiation in Weak Magnetic Fields," *Physical Review*, 24 (1924): 243–254.

33. Werner Heisenberg, "Quantum Theory and Its Interpretation," *Niels Bohr: His Life and Work*, pp. 94–108. The quotation is from p. 98.

34. *Archive for History of Quantum Physics*, Interview with W. Heisenberg on 13 February 1963, pp. 14–15.

35. Werner Heisenberg, "Quantum-Theoretical Re-Interpretation of Kinematic and Mechanical Relations," translated in *Sources of Quantum Mechanics*, pp. 261–276. (Originally published in *Zeitschrift für Physik*, 33 (1925): 879–893.)

36. Wolfgang Pauli, "On the Hydrogen Spectrum from the Standpoint of the New Quantum Mechanics," translated in *Sources of Quantum Mechanics*, pp. 387–415. (Originally published in *Zeitschrift für Physik*, 36 (1926): 336–363.) The quotation is from p. 387 of *Sources*; however, Van der Waerden's translation of the German *Veranschaulichung* as "visualization" is not correct.

37. Max Born, Werner Heisenberg, and Pascual Jordan, "On Quantum Mechanics II," translated in *Sources of Quantum Mechanics*, pp. 321–385. (Originally published in *Zeitschrift für Physik*, 35 (1926): 557–615.) The quotation is from p. 322 of *Sources*; however, Van der Waerden's translation of the German "anschauliche" as "visualizable" is not appropriate because anschauliche is an intuition that is superior to merely visualization. (See note 3.)

38. *Sources;* Here, too, Van der Waerden's translation of *anschaulich* as visualizable is not appropriate.

39. G. E. Uhlenbeck and S. Goudsmit, "Ersetzung der Hypothese vom unmechanischen Zwang durch eine Forderung bezüglich des inneren Verhaltens jedes einzelnen Elektrons," *Die Naturwissenschaften*, 13 (1925): 953–954 and Uhlenbeck and Goudsmit, "The Spinning Electron and the Theory of Spectra," *Nature*, 117 (1926): 264–265. For further discussions see Jammer, ch. 3 and B. L. van der Waerden, "Exclusion Principle and Spin," *Theoretical Physics in the Twentieth Century· A Memorial Volume to Wolfgang Pauli*, ed. M. Fierz and V. Weisskopf (New York: Interscience, 1960), pp. 199–244.

40. Max Born, *Problems of Atomic Dynamics* (Cambridge: The MIT Press, 1970), p. 128.

41. Andrade, *The Structure of the Atom*, p. 712.

42(a). Niels Bohr, "Über die Wirkung von Atomen bei Stössen," *Zeitschrift für Physik*, 34 (1925): 142–157; (b). Niels Bohr, "Atomic Theory and Mechanics," *Atomic Theory*, pp. 25–51. (See note 1.) (A version was published in *Nature* (Supplement) (5 December 1925).)

43. The two important experiments are: W. Bothe and H. Geiger, "Experimentelles zur Theorie von Bohr, Kramers und Slater," *Die Naturwissenschaften*, 13 (1925): 440–441 and A. H. Compton and A. W. Simon, "Directed Quanta of Scattered X-Rays," *Physical Review*, 26 (1925): 289–299.

44. ref. 42(a), p. 155. Unless indicated otherwise all translations are mine.

45. These researches are: Louis de Broglie, Thèse (Paris, 1924) and "Recherches sur la théorie des quanta," *Annales de Physique*, 3 (1925): 22–128; and Albert Einstein, "Quantentheorie des einatomigen idealen Gases," Berliner Berichte, (1924): 261–267; (1925): 3–14; (1925): 18–25. For further discussions of these papers see Jammer, ch. 5 and Klein (note 21).

46. ref. 42(b), p. 51.

47. Erwin Schrödinger, "Über das Verhältnis der Heisenberg-Born-Jordanschen Quantenmechanik zu der meinen," *Annalen der Physik*, 70 (1926): 734–756; portions are translated in *Wave Mechanics*, Gunther Ludwig, trans. (New York: Pergamon Press, 1968), pp. 127–150. The quotation is on p. 128 of the reprint volume. For further discussions of the relationship of Schrödinger's wave mechanics to the researches of de Broglie and Einstein see Klein (note 21) and Jammer ch. 5.

48. Erwin Schrödinger, "Der stetige Übergang von der Mikro-zur Makromechanik," *Die Naturwissenschaften*, 14 (1926): 664–666.

49. Werner Heisenberg, "Über den anschaulichen Inhalt der quantentheoretischen Kinematik und Mechanik," *Zeitschrift für Physik*, 43 (1927): 172–198, esp. pp. 184–189.

50. See Max Born and Pascual Jordan, "Zur Quantenmechanik," *Zeitschrift für Physik*, 34 (1926): 858–888, p. 879. (Translated in part in *Sources of Quantum Physics*, pp. 277–306; Van der Waerden's translation of "wahre Diskontinuumstheorie" (p. 300) as "essentially discontinuum theory" is not appropriate.)

51. See, for example, Jammer, ch. 5 and Klein (note 21).

52. Werner Heisenberg, "Mehrkörperproblem und Resonanz in der Quantenmechanik," *Zeitschrift für Physik*, 38 (1926): 411–426.

53. *Wave Mechanics*, p. 144 (see note 47).

54. Heisenberg, p. 412 (see note 52).

55. The principal papers are: (a) Max Born, "Zur Quantenmechanik der Stossvorgänge," *Zeitschrift für Physik*, 37 (1926): 863–867; (b) Max Born, "Quantenmechanik der Stossvorgänge," *Zeitschrift für Physik*, 38 (1926): 803–827, translated in part in *Wave Mechanics*, pp. 206–225; and (c) Max Born, "Physical Aspects of Quantum Mechanics," *Nature*, 119 (1927): 354–357.

56. ref. 55(a), p. 863.

57. ref. 55(b), p. 206 of *Wave Mechanics*.

58. ref. 55(a), p. 866.

59. ref. 55(c), p. 356.

60. Quoted from Jammer, p. 272.

61. *Archive for History of Quantum Physics*, Interview with W. Heisenberg on 22 February 1963, p. 30.

62. Werner Heisenberg, "Schwankungserscheinungen und Quantenmechanik," *Zeitschrift für Physik*, 40 (1926): 501–506.

63. ref. 61, p. 3. Indeed, Heisenberg's colleague Jordan rediscovered its results in a paper of 1927 and had to add a footnote that he had not been aware of Heisenberg's paper in ref. 62. See Pascual Jordan, "Über quantenmechanische Darstellung von Quantensprüngen," *Zeitschrift für Physik*, 40 (1927): 661–666, esp. p. 666.

64. Werner Heisenberg, "Quantenmechanik," *Die Naturwissenschaften*, 14 (1926): 889–994.

65. See *Archive for History of Quantum Physics*. Interview with W. Heisenberg on 25 February 1963, pp. 11–13.

66. "Discussion with Professor Werner Heisenberg," *The Nature of Scientific Discovery: A Symposium Commemorating the 500th Anniversary of the Birth of Nicolaus Copernicus*, ed. Owen Gingerich (Washington: Smithsonian Institution Press, 1975), pp. 556–573. The quotation is from p. 569.

67. ref. 18, pp. 286–287.

68. ref. 65, p. 18.

69. *Archive for History of Quantum Physics*, Interview with W. Heisenberg on 28 February 1963, p. 7. The relevant papers are: P.A.M. Dirac, "The Physical Interpretation of the Quantum Mechanics," *Proc. Roy. Soc. London* (A), 113 (1926): 621–641; Pascual Jordan, "Über eine neue Begründung der Quantenmechanik," *Zeitschrift für Physik*, 40 (1927): 809–838. These papers are discussed in Jammer, ch. 6.

70. *Archive for History of Quantum Physics*, Interview with W. Heisenberg on 5 July 1963, p. 11.

71. *Archive*, 27 February 1963, p. 27.

72. ref. 49, pp. 197–198.

73. A version of this lecture is in Niels Bohr, "The Quantum Postulate and the Recent Development of Atomic Theory," *Nature* (Supplement) (14 April 1928), 580–590; reprinted with a different introduction and no footnotes in ref. 1, pp. 52–91. For recent discussions of the genesis of Bohr's complementarity principle see: Jammer, ch. 7; Holton, "The Roots of Complementarity," in ref. 4, pp. 115–161; and K. Meyer-Abich, *Korrespondenz, Individualität und Komplementarität* (Wiesbaden: Steiner Verlag, 1965).

74. See, for example, Werner Heisenberg, *The Physical Principles of the Quantum Theory*, trans. C. Eckart and F. C. Hoyt (New York: Dover, 1963), "Preface." This text is from a set of lectures delivered at the University of Chicago in the spring of 1929.

75. For example, Werner Heisenberg, "Die Entwicklung der Quantentheorie, 1918–1928," *Die Naturwissenschaften*, 26 (1929): 490–496, esp. p. 494; *Archive for History of Quantum Physics*, Interview with W. Heisenberg on 12 July 1963, p. 6; and ref. 74, esp. pp. 157ff. The relevant papers are: P.A.M. Dirac, "The Quantum Theory of the Emission and Absorption of Radiation," *Proc. Roy. Soc. London* (A), 114 (1927): 243–265 (reprinted in J. Schwinger, ed., *Selected Papers on Quantum Electrodynamics* (New York: Dover, 1958), pp. 1–23); Pascual Jordan and Oskar Klein, "Zum Mehrkörperproblem der Quantentheorie," *Zeitschrift für Physik*, 45 (1927): 751–765; and Pascual Jordan and Eugene Wigner, "Über das Paulische Äquivalenzverbot," *Zeitschrift für Physik*, (1928): 631–651 (reproduced in Schwinger, pp. 41–61).

76. *Archive for History of Quantum Physics*, Interview with W. Heisenberg on 12 July 1963, p. 6.

Seymour Papert concurs with Poincaré that aesthetics rather than logic is the distinguishing feature of the mathematical mind. Papert refers to Poincaré's theory of mathematical creativity to reflect on the relationship between "the logical and extralogical in mathematics and the relationship between the mathematical and nonmathematical" in the spectrum of human activities.

He summarizes his essay as follows:

> Popular views of mathematics, including the one that informs mathematical education in our schools, exaggerate its logical face and devalue all connections with everything else in human experience. By so doing, they fail to recognize the resonances between mathematics and the total human being which are responsible for mathematical pleasure and beauty. This essay uses Poincaré's theory of mathematical thinking as an organizing center for reflections on the other face of mathematics, a face which is in touch with unconscious processes, with the individual's sense of being a person, and with the fact that the mathematician was first a child. Implicit in the confrontation of these views of mathematics is a broader question about the legitimacy of theories of psychology, often called cognitive, which seek to understand thinking in isolation from considerations of affect and aesthetics.

J.W.

THE MATHEMATICAL UNCONSCIOUS

SEYMOUR A. PAPERT

It is deeply embedded in our culture that the appreciation of mathematical beauty and the experience of mathematical pleasure are accessible only to a minority, perhaps a very small minority, of the human race. This belief is given the status of a theoretical principle by Henri Poincaré, who has to be respected not only as one of the seminal mathematical thinkers of the century but also as one of the most thoughtful writers on the epistemology of the mathematical sciences. Poincaré differs sharply from prevalent trends in cognitive and educational psychology in his view of what makes a mathematician. For Poincaré the distinguishing feature of the mathematical mind is not logical but aesthetic. He also believes, but this is a separate issue, that this aesthetic sense is innate: some people happen to be born with the faculty of developing an appreciation of mathematical beauty, and those are the ones who can become creative mathematicians. The others cannot.

This essay uses Poincaré's theory of mathematical creativity as an organizing center for reflections on the relationship between the logical and the extralogical in mathematics and on the relationship between the mathematical and the nonmathematical in the spectrum of human activities. The popular and the sophisticated wings of our culture almost unanimously draw these dichotomies in hard-edged lines. Poincaré's position is doubly interesting because in some ways he softens, and in some ways sharpens, these lines. They are softened when he attributes to the aesthetic an important functional role in mathematics. But the act of postulating a specifically mathematical aesthetic, and particularly an innate one, sharpens the separation between the mathematical and the nonmathematical. Is the mathematical aesthetic really different? Does it have common roots with other components of our aesthetic system? Does mathematical pleasure draw on its own pleasure principles or does it derive from those that animate other phases of human life? Does mathematical intuition differ from common sense in nature and form or only in content?

These questions are deep, complex, and ancient. My daring to address them in the space of a short essay is justified only because of certain simplifications. The first of these is a transformation of the questions similar in spirit to Jean Piaget's way of transforming philosophical questions into psychogenetic ones to which experimental investigations into how children think become refreshingly relevant. By so doing, he has frequently enraged or bewildered philosophers, but has enriched beyond measure the scientific study of mind. My transformation turns Poincaré's theory of the highest mathematical creativity into a more mundane but more manageable theory of ordinary mathematical (and possibly nonmathematical) thinking.

Bringing his theory down to earth in this way possibly runs the risk of abandoning what Poincaré himself might have considered to be most important. But it makes the theory more immediately relevant, perhaps even quite urgent, for psychologists, educators, and others. For example, if Poincaré's model turned out to contain elements of a true account of ordinary mathematical thinking, it could follow that mathematical education as practiced today is totally misguided and even self-defeating. If mathematical aesthetics gets any attention in the schools, it is as an epiphenomenon, an icing on the mathematical cake, rather than as the driving force which makes mathematical thinking function. Certainly the widely practiced theories of the psychology of mathematical development (such as Piaget's) totally ignore the aesthetic, or even the intuitive, and concentrate on structural analysis of the logical facet of mathematical thought.

The destructive consequences of contemporary mathematics teaching can also be seen as a minor paradox for Poincaré. The fact that schools, and our culture generally, are so far from being nurturant of nascent mathematical aesthetic sense in children causes Poincaré's major thesis about the importance of aesthetics to undermine his grounds for believing in his minor thesis which asserts the innateness of such sensibilities. If Poincaré is right about aesthetics, it becomes only too easy to see how the apparent rareness of mathematical talent could be explained without appeal to innateness.

These remarks are enough to suggest that the mundane transformation of Poincaré's theory might be a rich prize for educators even if it lost all touch with the processes at work in big mathematics. But perhaps we can have the best of both worlds. By adopting, as we shall, a more experiential mode of discussion through which theories about mathematical thinking can be immediately confronted with the reader's own mental processes, we do not, of course, renounce the possibility that the mathematical elite share similar experiences. On the contrary, that part of Poincaré's thinking which will emerge as most clearly valid in the ordinary context resonates strongly with modern trends which, in my view, constitute a paradigm shift in thinking about the foundations of mathematics. The concluding paragraphs of my essay will illustrate this resonance in the case of the Bourbaki theory of the structure of mathematics.

My goal here is not to advance a thesis with crisp formulations and rigorous argument, and it is certainly not to pass judgment on the correctness of Poincaré's theory. I shall be content (this is my second major simplification) to suggest to nonmathematical readers perceptions of, and a discourse about, mathematics which will place it closer than is commonly done to other experiences they know and enjoy. The major obstacle to doing so is a projection of mathematics which greatly exaggerates its logical face much as the Mercator projection of the globe exaggerates the polar regions so that on the map northern Greenland becomes more imposing than equatorial Brazil. Thus our discussion will polarize around separating and relating what I shall call the extralogical face of mathematics from the logical face. I shall ignore distinctions which ought to be made within these categories. Mathematical beauty, mathematical pleasure, and even mathematical intuition will be treated almost interchangeably insofar as they are representatives of the extralogical. And, on the other side, we shall not separate such very different facets of the logical as the formalists' emphasis on the deductive process, Russell's reductionist position (against which Poincaré fought so savagely), and Tarski's set theoretic semantics. These logical theories can be thrown together insofar as they have in common an intrinsic, autonomous view of mathematics. They deal with mathematics as self-contained, as justifying itself by formally defined (that is, mathematical) criteria of validity, and they ignore all reference of mathematics to anything outside itself. They certainly ignore phenomena of beauty and pleasure.

There is no theoretical tension in the fact that mathematical logicians ignore, as long as they do not deny, the extralogical. No one will call into question either the reality of the logical face of mathematics or the reality of mathematical beauty or pleasure. What Poincaré challenges is the possibility of understanding mathematical activity, the work of the mathematician solely, or even primarily, in logical terms without reference to the aesthetic. Thus his challenge is in the field of psychology, or the theory of mind, and, as such, has wider reverberations than the seemingly specialized problem of understanding mathematical thinking: his challenge calls in question the separation within psychology of cognitive functions, defined by opposition to considerations of affect, of feeling, of sense of beauty.

I shall, on the whole, side with Poincaré against the possibility of a "purely cognitive" theory of mathematical thinking but express reservations about the high degree of specificity he attributes to the mathematical. But first I must introduce another of the themes of Poincaré's theory. This is the role and the nature of the unconscious.

As the aesthetic vs. the logical leads us to confront Poincaré with the cognitive psychology, so the unconscious vs. the conscious leads to a confrontation with Freud. Poincaré is close to Freud in clearly postulating two minds (the conscious and the unconscious), each governed by its own dynamic laws, each able to carry out different functions with severely limited access to the other's activities. As we shall see, Poincaré is greatly impressed by the way in which the solution to a problem on which one has been working at an earlier time often comes into consciousness unannounced, and almost ready-made, as if produced by a hidden part of the mind. But Poincaré's unconscious is very different from Freud's. Far from being the site of prelogical, sexually charged, primary processes, it is rather like an emotionally neutral, supremely logical, combinatoric machine.

The confrontation of these images of the unconscious brings us back to our questions about the nature of mathematics itself. The logical view of mathematics is definitionally discorporate, detached from the body and molded only by an internal logic of purity and truth. Such a view would be concordant with Poincaré's neutral unconscious rather than with Freud's highly charged, instinct ridden dynamics. But Poincaré himself, as I have already remarked, rejects this view of mathematics; even if it could be maintained (which is already dubious) as an image of the finished mathematical product, it is totally inadequate as an account of the productive process through which mathematical truths and structures emerge. In its most naive form the logical image of mathematics is a deductive system in which new truths are derived from previously derived truths by means of rigorously reliable rules of inference. Although less naive logicist theses cannot be demolished quite so easily, it is relevant to notice the different ways in which this account of mathematics can be criticized. It is certainly incomplete since it fails to explain the process of choice determining how deductions are made and which of those made are pursued. It is misleading in that the rules of inference actually used by mathematicians would, if applied incautiously, quickly lead to contradictions and paradoxes. Finally, it is factually false as a description in that it provides no place for the as yet undebugged partial results with which the actual mathematician spends the most time. Mathematical work does not proceed along the narrow logical path of truth to truth to truth, but bravely or gropingly follows deviations through the surrounding marshland of propositions which are neither simply and wholly true nor simply and wholly false.

Workers in artificial intelligence have patched up the first of these areas of weakness, for example, by formalizing the process of setting and managing new problems as part of the work of solving a given one. But if the new problems and the rules for generating them are cast in logical terms, we see this as, at best, the replacement of a static logic by a dynamic one. It does not replace logic by something different. The question at issue here is whether even in the course of working on the most purely logical problem the mathematician evokes processes and sets problems which are not themselves purely logical.

The metaphor of wandering off the path of truth into surrounding marshlands has the merit, despite its looseness, of sharply stating a fundamental problem and preoccupation of Poincaré's: the problem of guidance, or, one might say, of "navigation in intellectual space." If we are content to churn out logical consequences, we would at least have the security of a safe process. In reality, according to Poincaré, the mathematician is guided by an aesthetic sense: in doing a job, the mathematician frequently has to work with propositions which are false to various degrees but does not have to consider any that offend a personal sense of mathematical beauty.

Poincaré's theory of how the aesthetic guides mathematical work divides the work into three stages. The first is a stage of deliberate conscious analysis. If the problem is difficult, the first stage will never, according to Poincaré, yield the solution. Its role is to create the elements out of which the solution will be constructed. A stage of unconscious work, which might appear to the mathematician as temporarily abandoning the task or leaving the problem to incubate, has to intervene. Poincaré postulates a mechanism for the incubation. The phenomenological view of abandonment is totally false. On the contrary, the problem has been turned over to a very active unconscious which relentlessly begins to combine the elements supplied to it by the first, conscious stage of the work. The unconscious mind is not assumed to have any remarkable powers except concentration, systematic operation, and imperviousness to boredom, distractions, or changes of goal. The product of the unconscious work is delivered back to the conscious mind at a moment which has no relation to what the latter is doing. This time the phenomenological view is even more misleading since the finished piece of work might appear in consciousness at the most surprising times, in apparent relation to quite fortuitous events.

How does the unconscious mind know what to pass back to the conscious mind? It is here where Poincaré sees the role of the aesthetic. He believes, as a matter of empirical observation, that ideas passed back are not necessarily correct solutions to the original problem. So he concludes that the unconscious is not able to rigorously determine whether an idea is correct. But the ideas passed up do always have the stamp of mathematical beauty. The function of the third stage of the work is to consciously and rigorously examine the results obtained from the unconscious. They might be accepted, modified, or rejected. In the last case the unconscious might once more be called into action. We observe that the model postulates a third agent in addition to the conscious and unconscious minds. This agent is somewhat akin to a Freudian censor; its job is to scan the changing kaleidoscope of unconscious patterns allowing only those which satisfy its aesthetic criteria to pass through the portal between the minds.

Poincaré is describing the highest level of mathematical creativity, and one cannot assume that more elementary mathematical work follows the same dynamic processes. But in our own striving toward a theory of mathematical thinking we should not assume the contrary either, and so it is encouraging to see even very limited structural resemblances between the process as described by Poincaré and patterns displayed by nonmathematicians whom we ask to work on mathematical problems in what has come, at MIT, to be called "Loud Thinking," a collection of techniques designed to elicit productive thought (often in domains, such as mathematics, they would normally avoid) and make as much of it as possible explicit. The example that follows illustrates aspects of what the very simplest kind of aesthetic guidance of thought might be. The subjects in the experiment clearly proceed by a combinatoric such as that which Poincaré postulates in his second stage until a result is obtained which is satisfactory on grounds that have at least as much claim to be called aesthetic as logical. The process does differ from Poincaré's description in that it remains on the conscious level. This could be reconciled with Poincaré's theory in many ways: one might argue that the number of combinatorial actions needed to generate the acceptable result is too small to require passing the problem to the unconscious level, or that these nonmathematicians lack the ability to do such work unconsciously. In any case, the point of the example (indeed, of this essay as a whole) is not to defend Poincaré in detail but to illustrate the concept of aesthetic guidance.

The problem on which the subjects were asked to work was the proof that the square root of 2 is irrational. The choice is particularly appropriate here because this theorem was selected by the English mathematician G. H. Hardy as a prime example of mathematical beauty, and consequently it is interesting, in the context of a nonelitist discussion of mathematical aesthetics, to discover that many people with very little mathematical knowledge are able to discover the proof if emotionally supportive working conditions encourage them to keep going despite mathematical reticence. The following paragraphs describe an episode through which almost all the subjects in our investigation pass. To project ourselves into this episode, let us suppose that we have set up the equation:

$\sqrt{2} = p/q$ where p and q are whole numbers

Let us also suppose that we do not really believe that 2 can be so expressed. To prove this, we seek to reveal something bizarre, in fact contradictory, behind the impenetrably innocent surface impression of the equation. We clearly have to do with an interplay of latent and manifest contents. What steps help in such cases?

Almost as if they had read Freud, many subjects engage in a process of mathematical "free association" trying in turn various transformations associated with equations of this sort. Those who are more sophisticated mathematically need a smaller number of tries, but none of the subjects seem to be guided by a prevision of where the work will go. Here are some examples of transformations in the order they were produced by one subject:

$$\sqrt{2} = p/q$$
$$\sqrt{2} \times q = p$$
$$p = \sqrt{2} \times q$$
$$(\sqrt{2})^2 = (p/q)^2$$
$$2 = p^2/q^2$$
$$p^2 = 2q^2$$

All subjects who have become more than very superficially involved in the problem show unmistakable signs of excitement and pleasure when they hit on the last equation. This pleasure is not dependent on knowing (at least consciously) where the process is leading. It happens before the subjects are able to say what they will do next, and, in fact, it happens even in cases where no further progress is made at all. And the reaction to $p^2 = 2q^2$ is not merely affective; once this has been seen, the subjects scarcely ever look back at any of the earlier transforms or even at the original equation. Thus there is something very special about $p^2 = 2q^2$. What is it? We first concentrate on the fact that it undoubtedly has a pleasurable charge and speculate about the sources of the charge. What is the role of pleasure mathematics?

Pleasure is, of course, often experienced in mathematical work, as if one were rewarding oneself when one achieves a desired goal after arduous struggle. But it is highly unplausible that this actual equation was anticipated here as a preset goal. If the pleasure was that of goal achievement, the goal was of a very different, less formal, I would say "more aesthetic" nature than the achievement of a particular equation. To know exactly what it is would require much more knowledge about the individual subjects than we can include here. It is certainly different from subject to subject and even multiply overdetermined in each subject. Some subjects explicitly set themselves the goal: "get rid of the square root." Other subjects did not seem explicitly to set themselves this goal but were nevertheless pleased to see the square root sign go away. Others, again, made no special reaction to the appearance of $2=p^2/q^2$ until this turned into $p^2 = 2q^2$. My suggestion is that the elimination of the root sign for the obvious simple instrumental purpose is only part of a more complex story: the event is resonant with several processes which might or might not be accessible to the conscious mind and might or might not be explicitly formulated as goals. I suggest, too, that some of these processes tap into other sources of pleasure, more specific and perhaps even more primitive than the generalized one of goal attainment. To make these suggestions more concrete, I shall give two examples of such pleasure giving processes.

The first example is best described in terms of the case frame type of calculus of situations characteristic of recent thinking in artificial intelligence. The original equation is formalized as a situation frame with case slots for "three actors," of which the principle or "subject" actor is $\sqrt{2}$. The other two actors, p and q, are subordinate dummy actors whose roles are merely to make assertions about the subject actor. When we turn the situation into $p^2 = 2q^2$, it is as sharply different as in a figure/ground reversal or the replacement of a screen by a face in an infant's perception of peek-a-boo. Now p has become the subject, and the previous subject, $\sqrt{2}$, has vanished. Does this draw on the pleasure sources that make infants universally enjoy peek-a-boo?

The other example of what might be pleasing in this process comes from the observation that 2 has not vanished away completely without trace. The 2 is still visible in $p^2 = 2q^2$! However, these two occurrences of 2 are so very different in role that identifying them gives the situation a quality of punning, or "condensation" at least somewhat like that which Freud sees as fundamental to the effectiveness of wit. The attractiveness and plausibility of this suggestion comes from the possibility of seeing condensation in very many mathematical situations. Indeed, the very central idea of abstract mathematics could be seen as condensation: the "abstract" description simultaneously signifies very different "concrete" things. Does this allow us to conjecture that mathematics shares more with jokes, dreams, and hysteria than is commonly recognized?

It is of course dangerous to go too far in the direction of presenting the merits of $p = 2q$ in isolation from its role in achieving the original purpose, which was not to titillate the mathematical pleasure senses but to prove that 2 is irrational. The statement of the previous two paragraphs needs to be melded with an understanding of how the work comes to focus on $p^2 = 2q^2$ through a process not totally independent of recognizing it as a subgoal of the supergoal of proving the theorem. How do we integrate the functional with the aesthetic? The simplest gesture in this direction for those who see the eminently functional subgoal system as the prime mover is to enlarge the universe of discourse in which subgoals can be formulated. Promoting a subordinate character (that is, p) on the problem scene to a principal role is, within an appropriate system of situation frames, as well-defined a subgoal as, say, finding the numerical solution of an equation. But we are now talking about goals which have lost their mathematical specificity and may be shared with nonmathematical situations of life or literature. Taken to its extreme, this line of thinking leads us to see mathematics, even in its detail, as an acting out of something else: the actors may be mathematical objects, but the plot is spelled out in other terms. Even in its less extreme forms this shows how the aesthetic and the functional can enter into a symbiotic relationship of, so to speak, mutual exploitation. The mathematically functional goal is achieved through a play of subgoals formulated in another, nonmathematical discourse, drawing on corresponding extramathematical knowledge. Thus the functional exploits the aesthetic. But to the extent we see (here in a very Freudian spirit) the mathematical process itself as acting out premathematical processes, the reverse is also true.

These speculations go some (very little) way toward showing how Poincaré's mathematical aesthetic sentinel could be reconciled with existing models of thinking to the enrichment of both. But the attempt to do so very sharply poses one fundamental question about the relationship between the functional and the aesthetic and hedonistic facets not only of mathematics but of all intellectual work. What is it about each of these that makes it able to serve the other? Is it not very strange that knowledge, or principles of appreciation, which is useful outside of mathematics, should have application within? The answer must lie in a genetic theory of mathematics. If we adopt a Platonic (or logical) view of mathematics as existing independently of any properties of the human mind, or of human activity, we are forced to see such interpretations as highly unlikely. In my remaining pages I shall touch on a few more examples of how mathematics can be seen from a perspective which makes its relationship to other human structures more natural. We begin by looking at another episode of the story about the square root of 2.

Our discussion of $p^2 = 2q^2$ was almost brutally nonteleological in that we discussed it from only one side, the side from which it came, pretending ignorance of where it is going. We now remedy this by seeing how it serves the original intention of the work which was to find a contradiction in the assumption 2= p/q. It happens that there are several paths one can take to this goal. Of these I shall contrast two which differ along a dimension one might call "gestalt vs. atomistic" or "aha-single-flash-insight vs. step-by-step reasoning." The step-by-step form is the more classical (it is attributed to Euclid himself) and proceeds in the following manner. We can read off from $p = 2q$ that p is even. It follows that p is even. By definition this means that p is twice some other whole number which we can call r. So

$$
\begin{aligned}
p &= 2r \\
p^2 &= 4r^2 \\
4r^2 &= 2q^2 \qquad \text{(remember: } p^2 = 2q^2\text{!)} \\
q^2 &= 2r^2
\end{aligned}
$$

and we deduce that q is also even. But this at last really is manifestly bizarre since we chose p and q in the first place and could, had we wished, have made sure that they had no common factor. So there is a contradiction.

Before commenting on the aesthetics of this process, we look at the "flash" version of the proof. It depends on having a certain perception of whole numbers, namely, as unique collections of prime factors: $6 = 3\times2$ and $36 = 3\times3\times2\times2$. If you solidly possess this frame for perceiving numbers, you probably have a sense of immediate perception of a perfect square (36 or p or q) as an even set. If you do not possess it, we might have to use step-by-step arguments (such as let $p = p_1p_2 \ldots p_k$, so that $p^2 = p_1p_1p_2p_2 \ldots p_kp_k$), and this proof then becomes, for you-here-and-now, even more atomistic and certainly less pleasing than the classical form. But if you do see (or train yourself to see) p and q as even sets, you will also see $p = 2q$ as making the absurd assertion that an even set (p) is equal to an odd set (q and one additional factor: 2). Thus, given the right frames for perceiving numbers, $p^2 = 2q^2$ is (or so it appears phenomenologically) directly perceived as absurd.

Although there is much to say about the comparative aesthetics of these two little proofs, I shall concentrate on just one facet of beauty and pleasure found by some subjects in our experiments. Many people are impressed by the brilliance of the second proof. But if this latter attracts by its cleverness and immediacy, it does not at all follow that the first loses by being (as I see it) essentially serial. On the contrary, there is something very powerful in the way one is captured and carried inexorably through the serial process. I do not merely mean that the proof is rhetorically compelling when presented well by another person, although this is an important factor in the spectator sport aspect of mathematics. I mean rather that you need very little mathematical knowledge for the steps to be forced moves, so that once you start on the track you will find that you generate the whole proof.

One can experience the process of inevitability in very different ways with very different kinds of affect. One can experience it as being taken over in a relationship of temporary submission. One can experience this as surrender to Mathematics, or to another person, or of one part of oneself to another. One can experience it not as submission but as the exercise of an exhilarating power. Any of these can be experienced as beautiful, as ugly, as pleasurable, as repulsive, or as frightening.

These remarks, although they remain at the surface of the phenomenon, suffice to cast serious doubt on Poincaré's reasons for believing that the faculty for mathematical aesthetic is inborn and independent of other components of the mind. They suggest too many ways in which factors of a kind Poincaré does not consider might, in principle, powerfully influence whether an individual finds mathematics beautiful or ugly and which kinds of mathematics he will particularly relish or revile. To see these factors a little more clearly, let us leave mathematics briefly to look at an example from a very sensitive work of fiction: Robert Pirsig's *Zen and the Art of Motorcycle Maintenance*. The book is a philosophical novel about different styles of thought. The principal character, who narrates the events, and his friend John Sutherland are on a motorcycling vacation which begins by riding from the east coast to Montana. Some time before the trip recounted in the book, John Sutherland had mentioned that his handlebars were slipping. The narrator soon decided that some shimming was necessary and proposed cutting shim stock from an aluminium beer can. "I thought this was pretty clever myself," he says, describing his surprise at Sutherland's reaction which brought the friendship close to rupture. To Sutherland the idea was far from clever; it was unspeakably offensive. The narrator explains: "I had had the nerve to propose repair of his new eighteen-hundred-dollar BMW, the pride of a half-century of German mechanical finesse, with a piece of old beer can!" But for the narrator there is no conflict; on the contrary: "beer can aluminum is soft and sticky as metals go. Perfect for the application . . . in other words any true German mechanic with half a century of mechanical finesse behind him, would have concluded that this particular solution to this particular technical problem was perfect." The difference proves to be unbridgeable and emotionally explosive. The friendship is saved only by a tacit agreement never again to discuss maintenance and repair of the motorcycles even though the two friends are close enough to one another and to their motorcycles to embark together on the long trip described in the book.

Sutherland's reaction would be without consequence for our

problem if it showed stupidity, ignorance, or an idiosyncratic quirk about ad hoc solutions to repair problems. But it goes deeper than any of these. Pirsig's accomplishment is to show us the coherence in many such incidents. This accomplishment is quite impressive. Pirsig presents us with materials so rich that we can use them to appreciate kinds of coherence implicit in them which are rather different from the one advanced by Pirsig himself. Here I want to touch briefly on two analogies between the story of Sutherland and the shim stock and issues we have discussed about mathematics: first, the relationship between aesthetics and logic in thinking about mathematics as well as motorcycles, and second, the lines of continuity and discontinuity between mathematics or motorcycles and everything else.

It is clear from the shim stock incident itself, and much more so from the rest of the book, that the continuity for Pirsig's characters between man, machine, and natural environment are very different and that these differences deeply affect their aesthetic appreciation. For the narrator, the motorcycle is continuous with the world not only of beer cans but more generally the world of metals (taken as substance). In this world, the metal's identity is not reducible to a particular embodiment of the metal in a motorcycle or in a beer can. Nor can any identity be reduced to a particular instance of it. For Sutherland, on the contrary, this continuity is not merely invisible, but he has a strong investment in maintaining the boundaries between what the narrator sees as superficial manifestations of the same substance.

For Sutherland, the motorcycle is not only in a world apart from beer cans; it is even in a world apart from other machines, a fact that enables him to relate without conflict to this piece of technology as a means to escape from technology. We could deepen the analysis of the investments of these two characters in their respective positions by noting their very different involvements in work and society. The narrator is part of industrial society (he works for a computer company) and is forced to seek his own identity (as he seeks the identity of metal) in a sense of his *substance* which lies beyond the particular form into which he has been molded. Like malleable metal, he is something beyond and perhaps better than the form which is now imposed on him. He certainly does not define himself as a writer of computer manuals. His friend Sutherland on the other hand is a musician and is much more able to take his work as that which structures his image of himself in the same way that he takes a motorcycle as a motorcycle and a beer can as a beer can.

We need not pursue these questions of essence and accident much further to make the important point, and a point which is widely ignored: if styles of involvement with motorcycle maintenance are intricated with such complexity with our psychological and social identities, one would scarcely expect this to be less true about the varieties of involvements of individuals with mathematics.

These ideas about the relationship of mathematical work with the whole person can be further illuminated by an example of an experiment in education, Turtle Geometry, as it is used with the LOGO programming language. These experiments express a critique of traditional school mathematics (which applies no less to the so-called new math than to the old). A description of this traditional mathematics in terms of the concepts we have developed in this essay would reveal it to be a caricature of mathematics in its depersonalized, purely logical, "formal" incarnation. Although we can document progress in the rhetoric of math teachers (teachers of the new math are taught to speak in terms of "understanding" and "discovery"), the problem remains because of what they are teaching.

In Turtle Geometry we create an environment in which the child's task is not to learn a set of formal rules but to develop sufficient insight into the way he moves in space to allow the transposition of this self-knowledge into programs that will cause a cybernetic animal, the turtle, to move. What is a Turtle? It can take several forms. It can be a machine controlled by a computer that crawls on the floor with a pen that leaves a trace of where it has been, or it can be a "light Turtle" which makes tracings on a computer display screen. The literature and ongoing research about the enterprise of developing Turtle Geometry and LOGO environments is more than we can summarize here, but what we want to do is underscore two closely related aspects of Turtle Geometry which are directly relevant to the concerns of this paper. The first is the development of an ego syntonic mathematics, indeed, of a body syntonic mathematics; the second is the development of a context for mathematical work where the aesthetic dimension (even in its narrowest of "the pretty") is continually placed in the forefront.

We shall give a single example which illuminates both of these aspects; an example of a typical problem that arises when a child is learning Turtle Geometry. The child has already learned how to command the Turtle to move forward in the direction that it is facing and to pivot around its axis, that is, to turn the number of degrees right or left that the child has commanded. With these commands the child has written programs which cause the Turtle to draw straight line figures. Sooner or later the child poses the question: "How can I make the Turtle draw a circle?" In Turtle Geometry we do not provide "answers." Learners are encouraged to use their own bodies to find a solution. The child begins to walk in circles and discovers how to make a circle by going forward a little and turning a little, by going forward a little and turning a little. Now the child knows how to make the Turtle draw a circle: simply give the Turtle the same commands one would give oneself. Expressing "go forward a little, turn a little" comes out in Turtle Language as REPEAT [FORWARD 1 RIGHT-TURN 1]. Thus we see a process of geometrical reasoning that is both ego syntonic and body syntonic. And once the child knows how to place circles on the screen with the speed of light, an unlimited palette of shapes, forms, and motion has been opened. Thus the discovery of the circle (and, of course, the curve) is a turning point in the child's ability to achieve a direct aesthetic experience through mathematics.

In the above paragraph it sounds as though ego syntonic mathematics was recently invented at MIT. This is certainly not the case and, indeed, would contradict the point made repeatedly in this essay that the mathematics of the mathematician is profoundly personal. It is also not the case that we have invented ego syntonic mathematics for children. We have merely given children a way to reappropriate what was theirs to begin. Most people feel that they have no "personal" involvement with mathematics, yet as children they constructed it for themselves. Jean Piaget's work on genetic epistemology teaches us that from the first days of life a child is engaged in an enterprise of extracting mathematical knowledge from the intersection of body with environment. The point is that, whether we intend it or not, the teaching of mathematics, as it is traditionally done in our schools, is a process by which we ask the child to forget the natural experience of mathematics in order to learn a new set of rules.

This same process of forgetting extralogical roots has until very recently dominated the formal history of mathematics in the academy. In the early part of the twentieth century, formal logic was seen as synonymous with the foundation of mathematics. Not until Bourbaki's structuralist theory appeared do we see an internal development in mathematics which opens mathematics up to "remembering" its genetic roots. This "remembering" was to put mathematics in the closest possible relationship to the development of research about how children construct their reality.

The consequences of these currents and those we encountered earlier from cognitive and dynamic psychology place the enterprise of understanding mathematics at the threshold of a new period heralded by Warren McCulloch's epigrammatic assertion that neither man nor mathematics can be fully grasped separately from the other. When asked what question would guide his scientific life, McCulloch answered: "What is a man so made that he can understand number and what is number so made that a man can understand it?"

References

1. Warren McCulloch, *Embodiments of Mind* (Cambridge: The MIT Press, 1965).

2. G. H. Hardy, *A Mathematician's Apology* (Cambridge: At the University Press, 1969).

3. R. Pirsig, *Zen and the Art of Motorcycle Maintenance: An Inquiry into Values* (New York: William Morrow and Co., 1974).

4. S. Papert, "Computers and Education," *The Study of the Future Impact of Computers on Information Processing*, Michael Dertouzos and Joel Moses, eds. (Cambridge: The MIT Press, to be published 1978).

5. S. A. Papert, "Uses of Computers to Advance Education," MIT Artificial Intelligence Laboratory, A-1-Memo #298 (Cambridge, 1973).

Howard Gruber's essay is concerned with the productive role played by complex images in scientific thinking. These images are often unexpressed in published writings, remaining a private source of inspiration. Gruber discusses "the *taxonomy* of imagination (images of perception, images of imagination, diagrammatic representations, concepts, etc.); the *plurality* of images (does a single great image ever capture a person's thought?); the *priority* of images in the growth of scientific ideas (do they generate ideas, underlie them, or merely express them *post hoc?*) and the *aesthetic* approach to science revealed in the use of such imagery."

The point of departure is Charles Darwin's recurrent image of the "irregularly branching tree" of nature, which antedated and foreshadowed the theory of evolution through natural selection. Aesthetically, it contrasted sharply with other systematic images of nature favored by Darwin's contemporaries and represented a rupture with a scientific past with its images of Platonic Perfection and celestial harmony. "The tree image suggests that we cannot predict everything, and more importantly, everything that might conceivably happen does not in fact do so. For Darwin, this accidental aspect of reality enhanced its value and its pathos."

J. W.

DARWIN'S "TREE OF NATURE" AND OTHER IMAGES OF WIDE SCOPE

HOWARD E. GRUBER

When we speak of the aesthetic attitude in science we have in mind aesthetic criteria that apply to the main results of scientific work, the perception of comprehensible order or pattern in some part of nature. We think of appropriate stories linking art and science in their appetite for pattern—a friend wishes to awaken the sleeping Mozart; he plays an unresolved progression of chords; Mozart jumps up, rushes to the piano, finishes the sequence. We resonate to the aesthetic motive behind Einstein's "God does not play dice with the world." (Not so different from the remark of an earlier physicist, Sir John Herschel, who complained that Darwin's theory of evolution was "the law of higgledy piggledy."[1]) Coupled with this aesthetic mood is a certain admiration for the heroic objectivity of scientists, their obstinate search for a place to stand from which to see into nature's order.

But behind these orderly results, which are after all our results and not nature's, lies nature itself, much wilder; and underneath them lies the often messy, inchoate processes of scientific thought. Is not our evident aesthetic pleasure in wild nature a part of the "aesthetics of science"? And are there not aesthetic feelings that apply, not only to the product but to the process of scientific work? And if so, do we invoke the same aesthetics of objectivity, simplicity, harmony, and order, or do we need another kind? Is there not also an aesthetic mood of erotic wildness, passionate involvement, pleasure in complexity and unpredictability? And if so, has this second aesthetic mood a place in science?

Others have made roughly similar distinctions: Herbert Read, the vital and the beautiful;[2] Alex Comfort, soft- and hard-centered;[3] Sylvano Arieti,[4] borrowing from Freud, primary and secondary processes; or simply romantic and classical. Such dichotomies are only approximations suited to their authors' purposes. Aesthetic experience and aesthetic process are many-faceted ensembles. They might submit to description by means of multivariate profiles. For the moment, then, I need make only a provisional and very rough dichotomy to help me get on with my argument.

Alex Comfort begins his essay, "Darwin and the Naked Lady," with a quotation from Paul Eluard, "*Rêve et réalité—la rose et le*

I thank Dr. Christiane Gillièron and Dr. William L. King for their helpful remarks on reading this paper and Dr. Ralph Culp for a stimulating preliminary discussion of Darwin's imagery.

rosier." Of course, Eluard meant *ideal* rather than *dream*, certainly not the fantasmagoric dreams of troubled sleep. In any event, for biological science the perfection of the rose and the tangle of the rose bush are both part of reality. If we want to grasp the aesthetic side of scientific work, it will not do to seize on one or the other. It is in their lively interplay that understanding moves forward. Since this interplay must inevitably take place within one person's experience, it is well to look steadily for a while at one person and to study these different kinds of aesthetic moments as they bear upon each other.

Among scientists, Charles Darwin is a useful subject. A critical period of his life were the years 1837–38; he was just back from the five-year voyage of the *Beagle*, he was twenty-eight years old, and he launched himself into a fifteen-month effort during which he constructed the essentials of the theory of evolution through natural selection. For this period he left us a set of revealing notebooks.[5] These permit us to see him in many moods at different junctures in the growth of his thought.

Darwin is a strategic choice for another reason. The meaning of his whole life work is saturated with the duality under discussion. On the one hand, he wanted to face squarely the entire panorama of changeful organic nature in its amazing variety, its numberless and beautiful contrivances, and its disturbing irregularity and imperfections. On the other hand, he was imbued with the spirit of Newtonian science and hoped to find in this shimmering network a few simple laws that might explain the whole movement of nature.

In the pages that follow I want to elaborate this theme and then illustrate it with examples drawn mainly from one generally neglected feature of scientific thought, the use of images of wide scope. These images are not usually part of the formal, consensus-minded part of science. They are more personal and therefore allow more room for the fruitful interplay of several aesthetic moods. I hope to draw together these two ideas, the interplay of different aesthetic moods in scientific thought, and the notion of images of wide scope.

On the Existence of Different Aesthetic Moods in Science

It seems hardly necessary to insist on the existence of the first mood, except to contrast it with its less often noticed companion. But a few reminders may be in order.

The search for an aesthetics of simple forms has an old history. As soon as the psychophysical methods were invented by Gustav Fechner, he applied them to the search for the proportions of the most pleasing rectangle. Although artists and designers have long known of the "golden section" (a rectangle whose proportions can be described by a simple mathematical formula), and although a presupposition in favor of it motivated early research, a general preference for it has not been easy to demonstrate. Nevertheless, the *idea* that there is some simple ratio describing the most pleasing proportions of any general form has been a seductive starting point for investigation.

Gestalt psychologists have emphasized perceptual tendencies toward closure, simplicity, and *prägnanz*. This last is hard to define but can be captured by the idea that we tend to see objects as being more like the ideal types they resemble than they really are. Thus, a slight departure from perfect circularity is still seen as a circle, etc.

When I was a student, we did not examine the aesthetic or other motives of scientists in any serious way. We were not asked to read, think, or write about the subject. It never occurred to me to doubt the occasional impassioned allusion to the connection between the search for beauty and the search for truth. Neither did it occur to me that there was any pressing need to study the matter: enough that I often felt a surge of pleasure at a pretty result or a beautiful idea.

But I did absorb a certain attitude toward the subject, one that fitted in pretty well with my own aesthetic preferences. What is beautiful in general and therefore beautiful in science is harmony, order, simplicity, a quality of cleanness. There was certainly not enough discussion of the subject to sort out two questions: Are we talking about beauty in nature or beauty in scientific work and thought? But if we had done so, we probably would have applied the same aesthetic criteria to both. The order of the scientific mind reflects the order of nature.

As time has gone by, the recognition of the importance of aesthetic values in scientific work has grown in me. At first this was an easy and welcome change, for the material I was drawn to happened to fit well with the clean aesthetics of simplicity. But more recently I have become increasingly aware that there is another interesting set of aesthetic attitudes. Things can be beautiful precisely because they are complex, unpredictable, imperfect, erotic.

Having begun with an admiration for Max Wertheimer's characterization of productive thinking as "fine, clean, direct,"[6] I was taken by surprise in my work on the growth of Darwin's thought to find that it was tortuous, tentative, enormously complex, full of unwarranted assumptions, and in a sense quite "dirty." At the same time, I saw that Darwin's picture of nature as an irregularly branching tree attributed to nature some of the characteristics that I saw in his thinking. One might sidestep a difficult aesthetic decision here by cleaving to "clean" aesthetics as it applies to the *products* of scientific thought; we could admit that the process is "dirty" while the product, such as Darwin's theory of evolution through natural selection, is clean and beautiful. Similarly, if we consider the scientists' view of beauty in nature itself, we might agree that nature is endlessly complex and aesthetically ambiguous, while beauty, residing in the eye of the beholder, is represented only by the simpler harmonies and patterns we can detect when we examine it in the quiet of the mind or the laboratory. Such distinctions may be useful if they help us to see that it is the concrete interaction among these different kinds of events that produce any particular aesthetic process.

No one reading the *Origin of Species*, especially the celebrated closing paragraph describing "the tangled bank,"[7] can fail to notice that Darwin took pleasure in the spectacle of complexity itself. And not only the complex entanglements of organisms at a moment in time, but the further manifold of intricacy residing in the meandering evolutionary path of every organism and every organ. Thus, on the side of the scientist's view of nature, here is at least one important figure whose image of nature itself is one of irregularity and entanglement. Moreover, he elaborates this image repeatedly over many years, with evident pleasure in ways that suggest an intimate connection between visual and poetic imagery and productive scientific thought.

When we consider the scientist thinking, we cannot escape the aesthetics of complexity. As we come to understand the intricacy of the course of thought, some of us admire it and find it all the more beautiful. As we see its unfinished character and the struggles of the scientist with a task which is inevitably and tragically beyond his grasp, other aesthetic values come to the fore. There is little prospect that our picture of creative thinking will grow simpler in the near future. We have just begun to uncover its seductive labyrinths.

But scientific thinking is not simply an object of investigation, it is also the lived experience of scientists. If in our role as spectators we can enjoy its wildness, so can we in our role as scientists, and so can our colleagues. Thus, the taking of pleasure in wildness is available to the whole intellectual community.

For a long time nothing so offended the aesthetic sensibilities of many scientists as the suggestion that the world was not perfectly orderly. When Herschel disdainfully described Darwin's theory as the "law of higgledy piddledy," this was not only an intellectual objection to the introduction of the idea of chance into a scientific theory but an aesthetic reaction as well. This is clear from Herschel's other remarks.[8]

But chance is a very broad concept that works in varying ways. A large number of similar events may produce a beautifully "simple" and predictable result, such as the smoothness of a Gaussian curve or the sphericity of a gas-filled balloon, and we may well admire such simplicity and regularity (either directly as a child does, or more sophisticatedly in taking note of both the simplicity and its underlying manifold). But the panorama of nature finds chance working in other ways where the results are neither simple, perfectly harmonious, or predictable. A real taxonomic tree has no simple order, and we must take our pleasures where we can—in our ability to make out its tortuous multiformity.

Darwin's Imagery

Recently I visited an exhibition of anamorphoses at the Brooklyn Museum: Distorted images are seen as normal when viewed from a special station point or when reflected in a cone or cylinder. One may see at the center of a picture a cone with a mirrored surface, surrounded by a distorted, unrecognizable image. At first, there is a tendency to glance perfunctorily at the distorted image, say of a human face or body, and then to study attentively the "corrected" version seen from the appropriate position. After a while, however, the distortions themselves draw the attention as objects of aesthetic interest. They have their own, sometimes weird, ugly, fascination. There is more to anamorphoses than a complex game of mapping a transformation. The artists who have played this game over the centuries are telling us something serious: nature has many faces, some harmonious and pretty, others wild and ugly.

It was precisely this duality that gave Darwin's contemporaries so much difficulty. Why would the Divine Artificer deliberately endow the natural order of His Creation with so much imperfection—hatred and violence, pestilence and death? How could these inescapable facts be reconciled with the image of a harmonious and perfect order of nature, the work of an omnipotent and benevolent Creator? There were various theological answers to the puzzle, and Darwin was thoroughly exposed to them in his university education. But when he begins his notebooks on evolution, we see from the first page that he has set himself the task of finding a completely natural solution to the dilemma. "Why is life short?" (*First Notebook*, p. 2). Why the cycle of birth, growth, reproduction, and death? His answer: to eliminate imperfections acquired in the life of the individual, ". . . generation destroys the effect of accidental injuries, which if animals lived for ever would be endless . . ." (p. 4). At the same time, the reproductive cycle permits adaptation: "There may be unknown difficulty with *full grown* individual with fixed organisation thus being modified,—therefore generation to adapt and alter the race to *changing* world." (p. 4, Darwin's italics). Thus, the function of the life cycle has a double aspect, to preserve the near-perfect adaptation already achieved, and to permit the organism to change when necessary?

From this vantage point Darwin moved quickly to his first theory of evolution, which I have described in detail elsewhere.[9] Monads or simple living forms arise through spontaneous generation; they evolve as they adapt to changing circumstances. Because of the fortuitous nature of their encounters with a changing world, their evolution takes the form of an irregularly branching tree. "Organized beings represent a tree, *irregularly branched*. . . . As many terminal buds dying as new ones generated. There is nothing stranger in death of species, than individuals." (p. 21, Darwin's italics). In quick succession, he makes three tree diagrams, each capturing somewhat different features of the idea that is growing in him. The first (fig. 1, upper diagram) emphasizes the idea of a triple branching: "Would there not be a triple branching in the tree of life owing to three elements—air, land and water, and the endeavour of each typical class to extend his domain into the other domains and subdivisions, three more, double arrangement. If each main stem of the tree is adapted for these three elements, there will be certainly points of affinity in each branch." (p. 24). This diagram and passage reflects certain general taxonomic problems Darwin was hoping to solve within the framework of his theory.

The second diagram (fig. 1, lower diagram) emphasizes the long gaps in the fossil record, a long dotted line showing a continuity between hypothetical extinct (but unknown) forms and the seemingly sudden efflorescence of a later group of organisms.

In the third diagram (fig. 2) Darwin introduces a specific notation to indicate a fundamental feature of his tree of life, extinction. Extinction was by no means a universally accepted idea, even among evolutionists. Lamarck had vigorously denied it. Darwin well knew that the fact of extinction is hard to prove, since it is founded on negative evidence—that is, the failure up to a particular moment in scientific time to find living specimens of an organism known to have once lived. Negative evidence is risky, counts for little. But Darwin had already begun to see that his tree image is a picture of exponential growth in the number of species, and this poses a problem for him that can only be solved by the idea of extinction. Thus he was at pains to show that extinction was not simply a fact but a formal requirement of his system. "I think Case must be that one generation then should have as many living as now. To do this and to have as many species in same genus *requires* extinction." (p. 36, Darwin's italics).

In short, the branching model, the image of the irregularly branching tree of nature played a pivotal role very early in his thinking about evolution. It captures many points: the fortuitousness of life, the irregularity of the panorama of nature, the explosiveness of growth and the necessity to bridle it "so as to keep number of species constant" (p. 37). And most important, the fundamental duality that at any time some must live and others die.

It took about fifteen months from this point until Darwin grasped the principle of natural selection as a key operator giving the tree of life its form. While Darwin's thought changed in many ways from these earliest notes until the time, some twenty years later, when he wrote the *Origin of Species*, this image of nature remained constant. Essentially the same tree diagram as his figure 3 appears in the *Origin*.[10] It is the only diagram in the book, and it is referred to throughout as he exploits its theoretical richness, some of which I have indicated.

Over the years, Darwin drew and redrew the tree diagram. I have paid attention to the scraps of paper in his manuscript on which these diagrams can be found, some dateable, others not. Some of them are hasty sketches, others painstakingly drawn and delicate traceries. On one such scrap there is the remark, "Tree not good simile—endless piece of seaweed dividing."[11] He is probably not so much correcting himself as searching for the right variant of his image to express a particular idea that has caught his attention, just as in the *First Notebook*, after his first drawing he wrote, "The tree of life should perhaps be called the coral of life, base of branches dead, so that passages cannot be seen" (p. 25).

We have seen how Darwin's view of the functional significance of the life cycle is connected with his panoramic view of nature as a whole. It is not often enough brought out that there was a certain cosmological cast to Darwin's thinking. Influenced, perhaps intimidated, by the empiricism of his day, Darwin later suggested that he worked in a "Baconian" fashion, inductively from facts to theory. His notebooks do not bear this out. He sketched his ideas with a broad brush and often drew a long bow. Thus, "Astronomers might formerly have said that God ordered each planet to move in its particular destiny. In same manner God orders each animal created with certain form in certain country, but how much more simple and sublime power let attraction act according to certain law, such are inevitable consequences—let animal be created then by the fixed laws of generation, such will be their successors. Let the powers of transportal be such, and so will be the forms of one country to another.—Let geological changes go at such a rate, so will be the number and distribution of the

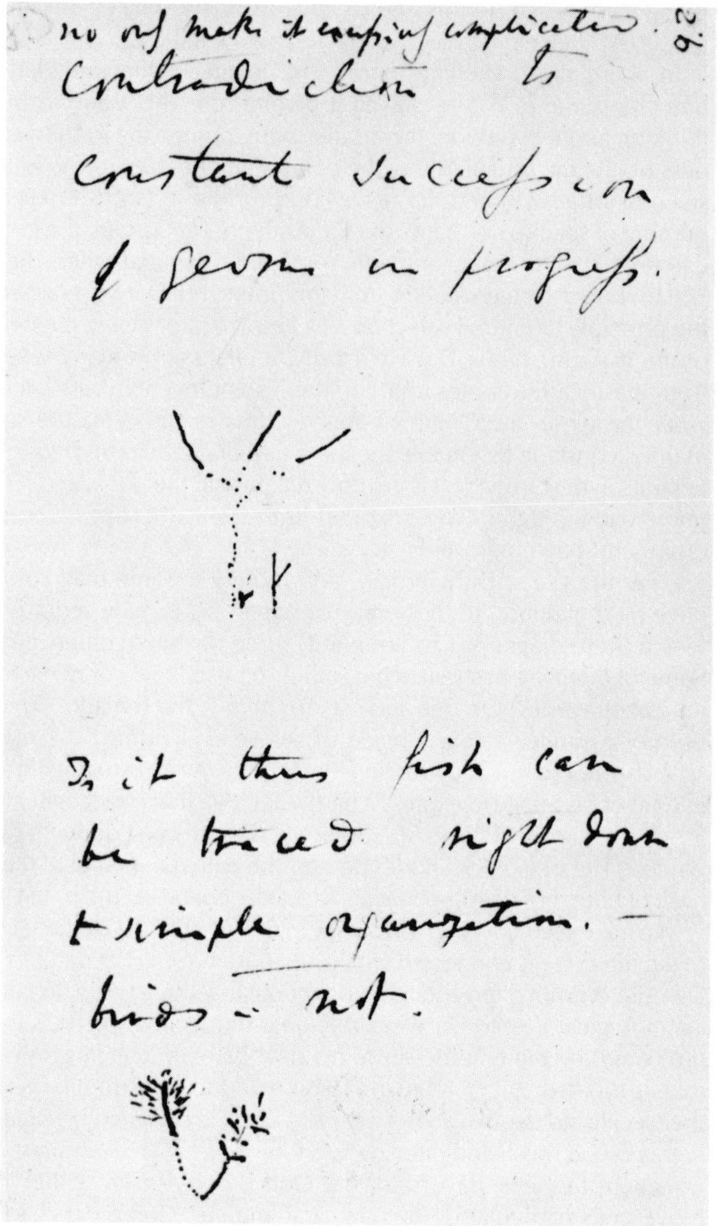

Figure 1. Darwin's first two tree diagrams, on page 26 of the *First Notebook*. Immediately preceding the upper tree the MS reads, "The tree of life should perhaps be called the coral of life, base of branches dead; so that passages cannot be seen.—[end of p. 25, beginning of p. 26] this again offers ((no only makes it excessively complicated)) contradiction to constant succession of germs in progress." Words in double parentheses were inserted above the line by Darwin.

Immediately preceding the lower tree the MS reads, "Is it thus fish can be traced right down to simple organization—birds—not." Courtesy of the Syndics of Cambridge University Library.

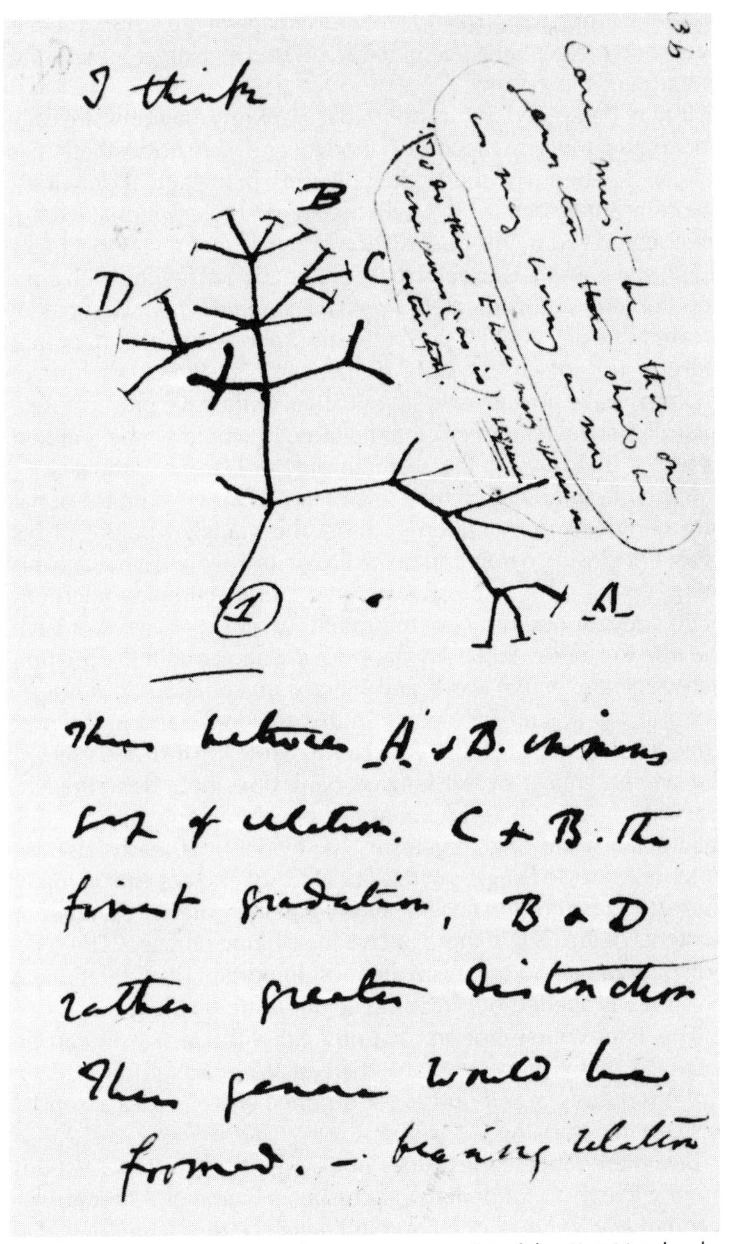

Figure 2. Darwin's third tree diagram, on page 36 of the *First Notebook*. The MS reads, "I think" followed by the diagram. Then, "Thus between A & B immense gap of relation, C & B, the finest gradation, B & D rather greater distinction. Thus genera would be formed,—bearing relation [end of p. 36, beginning of p. 37] to ancient types." The marginal insertion alongside the tree diagram reads, "Case must be that one generation then should have as many living as now. To do this & to have many species in same genus (as is), <u>requires</u> extinction." Courtesy of the Syndics of Cambridge University Library.

species!!" (pp. 101–102). In Darwin's image of the world, the life cycle and the evolving tree of life are nested in a larger view of the working of the cosmos.

It may be argued at this point that Darwin's diagrams are only conceptual tools for theoretical thought and have no aesthetic significance. Then why the evident pleasure in the actual drawings, the constant search for the right metaphor, the emotional excitement conveyed by his punctuation and frequent resort to a high-flown style? There is exactly that combination of feeling with concern for form and content that we have in mind when we speak of an aesthetic act. As well say that anamorphoses are not art, or that Dürer's use of instruments or Leonardo's studies of human anatomy have no aesthetic significance. Only if we presuppose a divorce between art and scientific thought would we be tempted to turn a blind eye to the aesthetic side of Darwin's imagery.[12]

If the irregularly branching tree of life is Darwin's image of nature deployed in evolutionary time, the "tangled bank" of his eloquent closing paragraph of the *Origin of Species* represents his image of the same explosive vitality in all its complex interconnectedness at one moment in time. It was this passage that gave the title to Stanley Edgar Hyman's lovely book about the relation between intellectual work and literary imagination.[13] Although foretastes of this passage occur in the notebooks and in the preliminary sketches of the *Origin* Darwin wrote in 1842 and 1844,[14] the precise image of the tangled bank does not. Nevertheless, Darwin's fascination with the intricate web of contemporaneous relationships among organisms is evident as early as the notebooks of the *Beagle* voyage (1831–36)[15] and appears in many forms long before the *Origin*. In the carefully drawn-up table of contents of the *Origin* there occurs the striking phrase, "The relation of organism to organism the most important of all relations." This idea is spelled out in some detail in the text.

This is of course not an idea original with Darwin. It can be found in many places, notably Gilbert White's *Natural History and Antiquities of Selbourne*.[16] It is nonetheless an idea essential to Darwin's thinking.

But interconnectedness does not by itself mean *imperfection*. Among Darwin's immediate precursors, in Lamarck's thought, in German *Naturphilosophie,* and in English Natural Theology, the idea of perfection was deeply embedded. In the first two it took the form of a scale of increasing perfection toward some limit or ideal type, that is, man, as in Lamarck's ladder of nature. The Natural Theologians could not accept this formulation because it meant that some of God's creation was less than perfect. In their view every organism was perfectly adapted to its place, seeming imperfections simply showed the limitations of our understanding of His work. Darwin's view of the natural order as inherently irregular, incomplete, and imperfect differed as radically from his predecessors' as did his view of the process by which this came about.

Thus two great and vital images, one of historical, the other of contemporary relations, form the substrate for the theory of evolution by natural selection. They are not, however, merely background, but are woven explictly into the theory as presented in the *Origin*.

The irregularly branching tree and the tangled bank represent the vital complexifying aspect of Darwin's thought. Other images must be sought that express the simplifying aspect. Of all those things that might occur in nature's incessant branching, some never do at all. The extinction of one evolutionary line makes impossible all the species that might have evolved from it. Of all those transient relations depicted in the tangled bank, some endure and others disappear. As we have already seen, the necessity for some principle of selection almost leaps out of these images.

In *Darwin on Man* I have carefully spelled out the slow process by which the thinker constructs the mental circumstances of his own insights. In considering even a most important moment of insight, this slow phase of construction must not be forgotten. On the other hand, such moments do occur and they deserve our attention. As is well known now, on September 25, 1838, Darwin, after reading (or perhaps rereading?) Malthus' *Essay on Population*, finally saw the principle of natural selection in a clear way. The passage in his notebook where he writes this out conveys the feeling of the moment as it was happening. It contains a striking and brutal mechanical image: "One may say there is a force like a hundred thousand wedges trying [to] force every kind of adapted structure into the gaps in the economy of nature, or rather forming gaps by thrusting out weaker ones" (*Third Notebook*, p. 135). The image reoccurs in the sketches of 1842 and 1844. In the first edition of the *Origin* it is heightened: "The face of Nature may be compared to a yielding surface, with ten thousand sharp wedges packed close together and driven inwards by incessant blows, sometimes one wedge being struck, and then another with greater force" (p. 67). This image has two features. On the one hand it emphasizes the idea that the theory deals with the interplay of a multitude of small forces rather than with the clash of Titans (in this way the theory is quite unlike Freud's depiction of struggle between Eros and Thanatos). On the other hand, the wedging image brings out the incessant rupturing of the seeming harmony of nature. Oddly, this image disappears from later editions of the *Origin*. It is as though Darwin needed to insist on this brutal rupturing but then recoiled a little from the hard mechanical nature of his image.

In the *Origin*, of course, two other well-known images occur. Human warfare, in Malthus' treatise the actual subject matter, becomes for Darwin one of the images he draws upon. For us the idea of warfare may seem to evoke the prodigal "wastefulness" of nature. For Darwin, there is a different emphasis. The image brings out the magnitude of the selection ratio of those that die to those that survive to reproduce. But Darwin's whole point is that this is not wasteful but creative, nature's way of fashioning the many ingenious contrivances embodied in every organism. Besides, if we avoid anthropocentrism, in the struggle for existence nothing is wasted. Those that die are eaten. Darwin is explicit on the way in which he intends the image of struggle to be taken: "I use the term Struggle for Existence in a large and metaphorical sense . . ." (*Origin*, p. 62)—covering many shades of meaning from actual combat between two organisms to simple dependence on conditions of life such as climate.

Finally, there is the metaphor of artificial selection. In one sense, Darwin's deep interest in artificial selection represents his desire to submit his theory to experimental test. Darwin was an inveterate experimentalist, and it must have troubled him that the theory as a whole could not be so tested. Human efforts to breed plants and animals come as close as possible. Darwin was, however, keenly aware of the many differences between artificial and natural selection, and his examination of the former, placed with poetic strategy in chapter 1 of the *Origin*, "Variation Under Domestication," has a clearly metaphoric intent. This metaphor plays a specific role in the theoretical structure, to emphasize the cumulative nature of evolutionary change. Darwin concludes the chapter by remarking that, of the several possible causes of change that he has discussed, natural selection is the most important: "Over all these causes of Change I am convinced that the accumulative action of Selection, whether applied methodically and more quickly, or unconsciously and more slowly, but more efficiently, is by far the predominant Power" (*Origin*, p. 43).

There are then at least these five images that Darwin used in developing the theory of evolution through natural selection—tree, tangled bank, wedging, war, and artificial selection. One of these, wedging, Darwin himself dropped from later editions of the *Origin*, so it is no wonder that it has been forgotten. Of the remaining four, only two are commonly referred to in discussing Darwin's theory—war and artificial selection. Both of these are simplifying images, dealing with the selective and corrective side of the process. The other two images, all too often forgotten, dramatize the principle of vitality, the explosive, irregular living material on which selection works.

For the moment, I simply want to draw attention to two points.

First, the multiplicity of these images. Second, the specificity of their functions in the theoretical structure. Earlier we saw that each drawing of the tree of life had specific features highlighting one or another aspect of the theory at an early stage of its development. Now the same can be said of Darwin's use of images in the definitive construction of his theory. They are not multiplied as a display of virtuosity but used with poetic economy, each image making its point, each point finding its image.

The Erotic Strain in Scientific Thought

Up to this point the discussion has been entirely cognitive: images generate ideas and ideas clarify images. But there is an affective aspect of the same work. We need to ask a new set of questions. Why did Darwin look so long and intently at nature in the first place? Why did he work so hard at science, care so much? It will not be necessary to go into unknown details of his early personality development to address the question of Darwin's emotional relation to science in some useful way.

As a result of certain lines of psychological research and a general readiness to accept the idea, it has become a widespread tendency to emphasize the role of theory in guiding scientific observation. "Observation is a theory-laden undertaking," insisted N. R. Hanson,[17] and Thomas Kuhn[18] has emphasized the same point. But one ought not slip carelessly from accepting this idea into believing that the connection between theory and observation is such that the former dictates the latter, for that would destroy the point of observation. Nor ought we slip into believing that an individual must have a clear or mature scientific theory before he can make valuable observations. The process of mapping data into scientific theory is a social process with many byways. It is possible, for example, for one person to make observations that another will explain.

The need for these cautionary remarks occurred to me from a simple fact in Darwin's life. He was a fine and eager observer *before* he was a great theoretician. The whole of his boyhood and adolescence at Shrewsbury, Edinburgh, and Cambridge attest to this. It is not only evident from his *Autobiography*[19] but from the still unpublished field notebooks he kept even before he went to Cambridge.[20] We need to account for the passion with which Darwin regarded nature without recourse to our knowledge of his later theoretical work. Something of the intensity of this passion is conveyed by the story of one of his entomological exploits: "But no pursuit at Cambridge was followed with nearly so much eagerness or gave me so much pleasure as collecting beetles. It was the mere passion for collecting, for I did not dissect them and rarely compared their external characters with published descriptions, but got them named anyhow. I will give a proof of my zeal: one day, on tearing off some old bark, I saw two rare beetles and seized one in each hand; then I saw a third and new kind, which I could not bear to lose, so that I popped the one which I held in my right hand into my mouth. Alas it ejected some intensely acrid fluid, which burnt my tongue so that I was forced to spit the beetle out, which was lost, as well as the third one" (*Autobiography*, p. 62). He was about eighteen years old then. Earlier at Edinburgh University he had taken part in the Plinian Society and made a few discoveries that found their way into print. Something energized young Darwin so that he looked at things of nature more often, more lengthily, and more intently—with greater passion—than the other young gentlemen of his acquaintance.

Carrying the story forward a little in time, during his last years at Cambridge and the early years of the voyage of the *Beagle*, Darwin was enamored of the writings of Alexander von Humboldt. At Cambridge he liked to read Humboldt's prose-poetic descriptions of nature aloud to his friends. On arriving in Brazil he wrote,

"Humboldt's glorious descriptions are & will for ever be unparalleled: but even he with his dark blue skies & the rare union of poetry with science which he so strongly displays when writing on tropical scenery, with all this fall far short of the truth. The delight one experiences in such times bewilders the mind; if the eye attempts to follow the flight of a gaudy butter-fly, it is arrested by some strange tree or fruit; if watching an insect one forgets it in the strange flower it is crawling over; if turning to admire the splendour of the scenery, the individual character of the foreground fixes the attention. The mind is a chaos of delight, out of which a world of future & more quiet pleasure will arise. I am at present fit only to read Humboldt; he like another sun illumines everything I behold."[21]

Here then is the "tangled bank" in an early form, almost a direct observation of it, not illuminated by a theory but by a poetic traveler. It is noteworthy that Darwin retained for so many years the capacity to enjoy both the "chaos of delight" and the quieter pleasure that followed from reflecting upon it.

Throughout his life, then—before he had any sort of theoretical position of his own to go on, before he had a "good" theory, and after—Darwin looked at nature with deep emotion. This was not disruptive but a positive force in his life. He saw well.

Among his earliest encounters with science was the reading of his grandfather's poetry. It is interesting that Dr. Erasmus Darwin was an evolutionist who propounded a theory something like Lamarck's. But it was probably more important for Charles Darwin that much of his grandfather's account of nature took a poetic form, and that much of this poetry was explicitly sexual. It is not today ranked as good poetry. In my own first reading of it, much of it seemed little better than doggerel. But as I read on, steeped myself in it, tried to feel it from young Darwin's point of view, I did not find it difficult to be moved, and to understand what a profound effect it might have had. It was poetry deliberately intended to excite the reader's passion for nature and for science, ranging from abstract ideas to close descriptions of sexual reproduction in plants. As befits poetry, the latter were cast in personified form, often so vivid that it is easy to forget that Dr. Darwin is not writing about human sexual behavior. Even in *Zoonomia*, the prose work in which he presented his theory of evolution, there is often a poetic ring. The chapter on instinct closes with the line, "Go proud reasoner, and call the worm thy sister!"[22] It was exactly this poetic admonition to which Charles Darwin devoted his life and his passion.[23]

Images of Wide Scope

We have seen that Darwin's thinking is characterized by the interplay of different aesthetic moods, and that the vehicle for much of his thought is a group of images of wide scope. An image is "wide" when it functions as a schema capable of assimilating to itself a wide range of perceptions, actions, ideas. This width depends in part on the metaphoric structure peculiar to the given image, in part on the intensity of the emotion which has been invested in it, that is, its *value* to the person.

Academic psychologists have contributed little to this subject. Most do not touch upon it at all. Even when dealing with the subject of imagery they limit themselves to representations of specific objects. But in trying to keep up with their colleagues in kindred sciences, they are coming closer to this topic. In Ulric Neisser's recent book, *Cognition and Reality*,[24] it is significant that in discussing this subject he must draw upon the work of sociologist Goffman, mathematician Minsky, architect Lynch, and anthropologist Gladwin. One of his two main examples is taken from Gladwin's *East Is a Big Bird*,[25] describing the way in which canoe navigators of the Puluwat Islands in the Pacific form an image permitting them to represent the relationships among stars, island landmarks, and ocean positions. His other example is taken from Lynch's book, *The Image of the City*,[26] representing the experience of a city by its inhabitants. Neither example is an image of the future or of the past or of an entirely imaginary world. (But Oscar Wilde wrote somewhere, "A map of the world that does not include Utopia is not worth even glancing at, for it leaves out the one country at which Humanity is always landing.")

For the most part the study of the role of images of wide scope in scientific thought has fallen between two stools. On the one hand, psychological examination of scientific thinking has been a process-oriented enquiry, aimed mainly at treating it as a variety of problem solving. The actual contents of scientific thought has not been deemed part of the real subject of investigation. Only recently, and in good part through the development of computer science, has it been recognized that ways of representing complex contents are inescapably part of the process too.

On the other hand, the psychological study of images has been mainly limited to images of specific objects. After a long period of disrepute, the notion of image has become newly respectable, and even prominent but only in a restricted way. For example, a person properly instructed can concoct combinations of images of specific objects linking the items in a series of terms to be remembered, thus greatly improving his performance in a recall task. But the study of large and complex images has not yet been taken up with any vigor.

Yet it is almost obvious that the contents and therefore the process of scientific thought include a great deal of imagery. Many of the images in question represent things unseen and unfelt, and some of them represent things which, in the thinker's own conception of them, are nonexistent. Although psychologists have neglected the subject, Kenneth Boulding in his book, *The Image*, wrote, "It is the capacity for organizing information into large and complex images which is the chief glory of our species."[27] Although Boulding speaks of "information" he does not mean to restrict his remark to images of the real world. It is part of his thinking about the subject that the human stock of images includes selfconscious information about the products of imaginative construction.

What is the function of such complex imagery? Boulding, Miller, Galanter, and Pribram,[28] Minsky,[29] and others have treated them as indispensable for the activity of ordinary life. Information is organized in complex packages, schemas, or frames, and these are activated as needed. New perceptual data are mapped into them, behavior is regulated by them. Each person probably possesses a very large stock of such images that can be flexibly recombined with each other.

But for our present purposes we must select from among this wealth a smaller set of images. These are the ones that are deliberately chosen to carry the special message the individual scientist is trying to formulate and convey. They are not simply chosen, but constructed, winnowed out, criticized, and reconstructed. They are the product of hard, imaginative, and reflective work, and in their turn they regulate the future course of that work. The scientist needs them in order to comprehend what is known and to guide the search for what is not yet known. He must represent to himself the possible unknowns. But these representations are not fragmentary bits of speculation. Just as we need organized schemas to represent the world as it is, we need them to represent the world as it might be. The special, empirically testable hypotheses scientists sometimes construct in a seemingly neutral spirit are really sample products of the activity of some organized representation of the world. The movement, from such hypotheses to a conscious delineation of their source-images, marks an important turning point in the growth of any scientific thought process.

Such images lend palpability to otherwise vague ideas. This feature accounts for the ability of images of wide scope to liberate the kind of high excitement that permits prolonged and attentive labor, so evident in scientific work. We most often see this excitement displayed when the thinker focuses attention upon a single great image. This focus gives rise to Illusions of Monolithicity. We are tempted to look for the one great image that motivates an individual or the members of a discipline, or of an age. Even when we recognize that an intricate theory is formed in the interplay of ideas, there remains a strong tendency to summarize the individual as though there is an allotment of one great image to each great person.

But as I have tried to suggest, not only is there a plurality of images, but this is a necessary condition for fruitful work. Although we do not know anything about how many complex images a person can or ought to work with, a little guesswork may not hurt. If we consider Darwin's image of the tree of nature, we can see that it took many hours of work to fashion and refashion it. It reflected long study of many special taxonomic problems and alternative taxonomic schemes. Meanwhile, in another domain, plant and animal breeding, a great labor was necessary for Darwin to reach the point where he could see the useful analogy between artificial and natural sleection. And so on, for each of his enterprises. Useful images of wide scope do not come as cheaply as the ideas thrown out hastily in a brainstorming session (imagine Einstein trying to brainstorm!). Moreover, to use them well—to examine the intricacies of each one, to find new and fruitful combinations of them, to express these ideas understandably—all this takes its toll of time and energy. It would be my guess then that the number of images of wide scope that one individual can bring to bear in one lifetime of scientific work is rather small, something like fifty or one hundred, that is, not one and not many thousands. Perhaps the individual can cope with 4–5 wide images that serve as leitmotifs for an entire life and a somewhat larger number, say 50–100, that are used in the elaboration of these thematic organizers. How small this number is can be felt, if not seen, by comparing it with the number of images unselected subjects can produce in a mnemonic processing experiment of the sort I referred to earlier. In one such experiment I found that subjects could produce about one useful pair of "narrow" images every five seconds, a rate of about 600 in a 50-minute hour.[30]

The issue of number has a bearing on what we may call the erotic side of scientific work. In some general sense, every scientist may form an emotional cathexis with the whole of nature, or better, with his *idea* of the whole. But in actual work we see that every person must make severe choices. This is not merely a matter of the time it takes to get the work done or to learn enough to do it. Much of the time goes into forming a deep enough cathexis with some particular set of natural objects or ideas to permit the steady engagement of the person's whole effort. Such love is not formed in an instant. In matters of work the scientist may be polygamous but not promiscuous. Creativity demands commitment.

The need for commitment can be turned around and looked at in another way. One of the functions of complex images is to give the thinker something almost palpable to permit the formation of a productive cathexis. Not only people and animals but ideas can be lovable (or hateful). And just as students of literature tell us that it is not quite the person but our image of him that we love, it may well be that it is not quite the idea or concept that we love but the image from which the idea is formed.

This attachment, with its attendant access to the person's whole value system (which Boulding treats as another set of images), may help to explain an otherwise quite puzzling experience. When we hear of a new idea or a new finding, we often know with a sense of great immediacy that it "feels right" or that it "feels wrong." Only later do we work out our reasons.

I began this discussion by distinguishing two aesthetic moods in science and suggesting that productive scientific work depends on a lively interplay between them. But these two moods each have many features, and we may expect that every individual scientist will have a different aesthetic profile. I have used the examination of Darwin's complex imagery as a vehicle for exploring these two moods. The subject of such imagery remains quite dark, deserves study. Although some images are shared, many are personal. They are so personal that even when the individual displays them openly they may go unnoticed. This is a gloomy picture. Can we only communicate successfully by means of highly simplified sketches of the products of thought? Is the intimacy of sharing thought itself and the feelings that go with it—these most human of all experiences—is this beyond our reach? Perhaps so. But perhaps we have merely not yet lived in a world where thinking men and women really stop to listen to each other or to take long and loving looks at each other's images. Is this impossible?

Notes

1. Charles Darwin, *The Life and Letters of Charles Darwin*, ed. Francis Darwin, 3 vols. (London: Murray, 1887), II, p. 241.

2. Herbert Read, *Icon and Idea: The Function of Art in the Development of Human Consciousness* (Cambridge: Harvard University Press, 1955).

3. Alex Comfort, *Darwin and the Naked Lady: Discursive Essays on Biology and Art* (New York: George Braziller, 1962).

4. Silvano Arieti, *Creativity: The Magic Synthesis* (New York: Basic Books, 1976).

5. There are two sets of notebooks, those on evolution per se and those on associated issues concerning man, mind, and materialism. "Darwin's Notebooks on Transmutation of Species," *Bulletin of the British Museum (Natural History)*, Historical Series, vol. 2, nos. 2–6, 1960–61; vol. 3, no. 5, 1967, ed. Sir Gavin de Beer, M. J. Rowlands, and B. M. Skramovsky. These are referred to in the text as *First Notebook*, *Second Notebook*, etc. The notebooks on man, etc. are published in full in Howard E. Gruber, *Darwin on Man: A Psychological Study of Scientific Creativity* together with *Darwin's Early and Unpublished Notebooks*, ed. Paul H. Barrett (New York: E. P. Dutton, 1974).

There was also a field notebook, mainly on geological matters but with a few observations on animal breeding, that Darwin kept during 1838. No transcription of this has yet been published, but it is described in Rudnick's excellent account of Darwin's struggle with a major geological puzzle that had a bearing on his general point of view: the peculiar "parallel roads of Glen Roy," which led Darwin to make a field trip to Scotland in July of 1838, the same month that he began the notebooks of man, mind, and materialism. See Martin Rudnick, "Darwin and Glen Roy: A 'Great Failure' in Scientific Method?" *Studies in the History and Philosophy of Science*, 5 (1974), no. 2, pp. 97–185.

6. Max Wertheimer, *Productive Thinking* (New York: Harper & Row, 1945), p. 189.

7. Charles Darwin, *On the Origin of Species by Means of Natural Selection, or the Preservation of Favoured Races in the Struggle for Life* (London: Murray, 1859), pp. 489–490.

8. Charles Darwin, *More Letters of Charles Darwin*, ed. Francis Darwin and A. C. Seward, 2 vols. (London: Murray, 1903), I, pp. 190–191.

9. Gruber and Barrett, pp. 129–149.

10. *Origin*, p. 116.

11. Darwin Mss, Cambridge University Library.

12. A number of writers have discussed the relation between schematization and art, none of them suggesting a dichotomy. See Arieti, *Creativity*, Read, *Icon and Idea*, and Rudolf Arnheim, *Visual Thinking* (Berkeley: University of California Press, 1969), and E. H. Gombrich, *Art and Illusion: A Study in the Psychology of Pictorial Representation* (New York: Pantheon Books, 1960).

13. Stanley Edgar Hyman, *The Tangled Bank: Darwin, Marx, Frazer and Freud as Imaginative Writers* (New York: Atheneum, 1962).

14. Charles Darwin, *The Foundations of the Origin of Species: Two Essays Written in 1842 and 1844*, ed. Francis Darwin (Cambridge: At the University Press, 1909).

15. See H. E. Gruber and Valmai Gruber, "The Eye of Reason: Darwin's Development During the *Beagle* Voyage," *Isis*, 53 (1962):186–200.

16. Gilbert White, *Natural History and Antiquities of Selborne*, 1789.

17. Norwood Russell Hanson, *Patterns of Discovery* (Cambridge: At the University Press, 1958).

18. Thomas S. Kuhn, *The Structure of Scientific Revolutions* (Chicago: University of Chicago Press, 1962).

19. Charles Darwin, *The Autobiography of Charles Darwin 1809–1882: With Original Omissions Restored*, ed. Nora Barlow (London: Collins, 1958).

20. Darwin Mss, Cambridge University Library.

21. Charles Darwin, *Charles Darwin's Diary of the Voyage of H. M. S. Beagle*, ed. Nora Barlow (Cambridge: At the University Press, 1934).

22. Erasmus Darwin, *Zoonomia; or The Laws of Organic Life*, 2 vols. (Dublin: Byrne, 1800), vol. I, p. 219.

23. There is much more that could be said of Darwin and poetry. During the voyage he carried Milton's *Paradise Lost* in his pocket on his expeditions ashore. No one has traced out the connections between Darwin and Milton in any detail. A full work on the poetic vein in Darwin's thought remains to be written.

24. Ulric Neisser, *Cognition and Reality* (San Francisco: W. H. Freeman, 1976).

25. Thomas Gladwin, *East is a Big Bird* (Cambridge: Harvard University Press, 1970).

26. Kevin Lynch, *The Image of the City* (Cambridge: The MIT Press, 1960).

27. Kenneth Boulding, *The Image* (Ann Arbor: University of Michigan Press, 1956).

28. George A. Miller, Eugene Galanter, and Karl H. Pribram, *Plans and the Structure of Behavior* (New York: Holt, Rinehart, and Winston, 1960).

29. Marvin Minsky, "A framework for representing knowledge," *The Psychology of Computer Vision*, P. H. Winston, ed. (New York: McGraw-Hill, 1975).

30. H. E. Gruber, A. Kulkin, and P. Schwartz, "The effect of exposure time on mnemonic pressing in paired associate learning," Paper read at Eastern Psychological Association, 1965.

Sir Geoffrey argues that all knowing is based on a form of conceptual architecture and depends, like other skillful doing, on the ability to impose, recognize, and combine forms. He distinguishes this mental capacity from the capacity for logical deduction and analysis and argues that it is underrated and even ignored only because of our unwillingness to admit the existence of mental activities which we cannot fully describe, a reluctance which is itself a product of recent Western culture. He follows this line of thought from the simplest forms of perception and cognition to deliberate physical and social design.

J. W.

RATIONALITY AND INTUITION

GEOFFREY VICKERS

The Causal and the Contextual

Why not "aesthetics in science"? Whence comes the implication that to find aesthetics in science is like finding poetry in a timetable? The answer lies in the sad history of Western culture which, over the last two centuries, has so narrowed the concepts of both Science and Art as to leave them diminished and incommensurable rivals—the one an island in the sea of knowledge not certified as science; the other an island in the sea of skill not certified as Art.

This debasement is relatively new. In medieval universities all fields of knowledge open to organized study were scientiae, and all fields of skill open to organized acquisition were Arts. Rhetoric and astronomy were equally scientiae; but the title accorded to the student who satisfied his examiners in these and the other recognized scientiae was that of a Master of Arts. Similarly, Art was not separated from technology; there was an Art and Mystery of Bricklaying. Cellini made a splendid pair of front doors. Where was the boundary between art and architecture, mason, carver, builder, and architect?

Moreover the two words "Ars" and "Scientiae" not only embraced virtually all skill and knowledge, but also overlapped each other's territory without offense. Everyone knew that knowing was a skilled *activity*, an *Art*. Both words connoted both product and process—on the one hand an accumulating store of knowledge and artifacts; on the other hand a growing heritage of transmitted skills and standards of skill and excellence in knowing and doing.

Science acquired its present limited meaning barely before the nineteenth century. It came to apply to a method of testing hypotheses about the natural world by observations or experiments which might give results inconsistent with the hypothesis to be tested. Thence it came to comprise the growing body of related hypotheses which had survived these tests. The method never explained wholly and often failed to explain at all how the hypothesis originally emerged. But this fact was not generally ac-

This paper is a greatly extended version of a lecture which I gave at the Massachusetts Institute of Technology in April, 1974. It embodies and expands parts of a longer unpublished paper called "The Tacit Norm," prepared for a symposium on The Moral and Esthetic Structure of Human Adaptation held by the Wenner-Gren Foundation for Anthropological Research at Burg Wartenstein in July, 1969 on which I drew for the original lecture.

knowledged until this century. Even now it courts opposition to describe a scientific theory as a work of *art*, largely because of the corresponding narrowing of the concept of Art.

Yet few would deny that a scientific hypothesis, a technological invention, a plan for a new city, a painting, a musical composition, and a new law are all human *artifacts*, skillful making by human minds of designs for ordering or explaining some aspect of what we experience as reality. And few would deny that all such designing involves the creation, imposition, and recognition of *form*.

Equally, few would deny that particular achievements such as these are episodes in a process of change which proceeds continuously, though slowly and often unconsciously, for example, in the kind of explanation which scientific minds find acceptable, in the kind of methods by which technologists approach their problems, in the aesthetic idiom in which artists express themselves, and in the ethical standards which lead societies to change their laws and customs. History reveals in retrospect the presence of standards in all these fields which guide those who work in them and those who criticize their work and which are themselves changed both by the creations which they guide and by the controversies which they provoke.

I am not denying in the least that there are differences between the different fields in which these "arts" are practiced. Indeed, if I had space to pursue the theme, I would insist that these differences are much greater than are usually admitted either by scientists or by any of those who use the word "science" as a generic term. I would insist that different fields of possible knowledge (scientiae) admit such different kinds as well as degrees of knowledge that it is confusing to class as "science" even all those fields which aspire to the name. There are important differences between the natural sciences and the logical sciences which include all the branches of mathematics and symbolic logic. There are even more striking differences between the natural sciences and what Herbert Simon[1] has called the sciences of the artificial, by which he means the fields of knowledge of which the subject matter is partly man-made. Virtually our whole environment, he insists, is partly artificial in this sense. Not only tools, machines, and buildings but also institutions, languages, and cultures are human artifacts. What "scientific" knowledge, he asks, is possible about a subject matter which might be other than it is?

I do not find his answer adequate, even on a very limited definition of "science," but I warmly approve of the distinction which he draws. The regularities to be found in the "artificial" world are different in origin, kind, and reliability from those to be found in the natural world. The "laws" of England are not "laws of nature," and we have access to different means of knowing about them, notably a knowledge of human history which is open to us only because we ourselves are human.

I do not propose to pursue these differences here because I am concerned to explore the mental processes which are common to them all and especially the element connoted by the word aesthetics in the title to this book. My thesis is that the human mind has available to it at least two different modes of knowing and that it uses both in appropriate or inappropriate combinations in its endless efforts to understand the world in which it finds itself, including its fellow human beings and itself. One of these modes is more dependent on analysis, logical reasoning, calculation, and explicit description. The other is more dependent on synthesis and the recognition of pattern, context and the multiple possible relations of figure and ground. The first involves the abstraction and manipulation of elements, irrespective of the forms in which they are combined. The other involves the recognition or creation of form, irrespective of the elements which compose it. Both are normal aspects of the neocortical development which distinguishes man from his fellow mammals. Both are needed and both are used in most normal mental operations.

They are often referred to as rationality and intuition, and the names would serve as well as any other, were it not that a difference in the character and function of the two capacities has attached to intuition, in our contemporary culture, a load of mischievous and misleading connotation.

The main difference to which I refer is that a rational process is fully describable, whereas an intuitive process is not. Because our culture has somehow generated the unsupported and improbable belief that everything real must be fully describable, it is unwilling to acknowledge the existence of intuition; and where it cannot avoid doing so, it tends to confine it to the area where the creative process is least constrained and most in evidence—namely the narrow contemporary concept of Art—so much so that when this ubiquitous faculty appears in the practice of "science," it is greeted as a strange incursion from a foreign field called "aesthetics." But in my view this approach half accepts the cultural confusion which I wish to contest.

The theory of biological evolution is a convenient example. For a century before Darwin and Wallace the fact of biological evolution had been forcing itself into the consciousness of Western man. It was opposed by the strange belief, accepted for more than a thousand years, that each and every *possible* biological form had been specially created by a divine demiurge so obsessively creative that he could not leave any conceivable form unrealized. Lovejoy[2] has documented, in fascinating detail, the history of this theory and its eventual decay.

The main agent in its decay was the discovery of the fossil record. Here in sufficiently exact chronological order was a sequence of biological forms which exhibited continuity and discontinuity through change with time. Eohippus was a far cry from the favorite contemporary racehorse, yet the development of one from the other was clear enough.

Why was it clear? Measurement played no significant part in these acts of recognition. They were exercises in the human capacity of appreciating, comparing, and contrasting *form*. They threw no light on how these developments took place or why some died out. That had to await a *theory*. But the apparent fact arrested human attention before there was a theory to explain it and provided the driving power to seek a theory. Without it there would have been no theory of evolution, for there would have been nothing to explain.

This intuitive sense of form entered also into the theory, the explication, no less than into the explicandum. The theory of natural selection implied a theory of particulate inheritance which did not exist in Darwin's day and would not exist for fifty years; indeed it seemed inconsistent with the view of biological inheritance then currently held. For if all inherited traits were mixed in a kind of general broth, no advantageous element could have survived long enough to develop its potential. Darwin was troubled when this was pointed out to him. His intuitive grasp of the way biological inheritance must work felt right. But ought he to trust it before theory had established its rightness by propounding and testing an explanation of *how* it worked? Happily he was already too deeply committed to withdraw. Time was to justify him.[3]

Are we to identify Darwin's intuition about the way inheritance must work with the intuition which asserted the fact of biological evolution? I think we should. If it strains the concept of "aesthetics," even in the wide sense used in this book, let us widen that concept still further or choose some other term with less constraining implications.

It is, of course, perfectly reasonable to *mistrust* a faculty which is not fully describable (even though we cannot do without it), since its obscurity makes it hard to verify. We should expect then that the main function of the rational process would be to critize and test so far as it can the products of the intuitive process. And this is, of course, precisely what it does, as the history of science so clearly shows. It can do so only in varying degrees and the less it can do so the more trust we have to repose in other tests by which we come to accept or to change the product of our intuition.

The history of the natural sciences is full of once accepted intuitions (such as the "ether") which were later found to be unnecessary or wrong. Some of them held up the progress of science for centuries. Such was the intuitive belief that the paths of the planets must be circular. Some proved useful though wrong. Such was the original atomic theory. For two thousand years science proceeded on the unverified assumption that all material forms must be constructed of basic individual elements, capable of countless combinations but not themselves further divisible. Within a few years after the existence of atoms was first actually demonstrated, it was found that they lacked both the characteristics with which they had been credited. They were neither indivisible nor indestructible. Yet atomic theory, so far from receiving its death blow, took off into the new world of subatomic physics. One of the first dividends was an understanding of at least some of the forces which enabled atoms to combine, a fact predicated in the original theory, yet wholly inexplicable if atoms were, in fact, no more than elemental billiard balls.

I have therefore avoided so far as possible in this paper the use of the expression "science and aesthetics." I have chosen instead to concentrate on the relation between rationality and intuition. This is, indeed, well exemplified in the recent development of physics and mathematics to which this book is largely directed. But it is not confined to that context. I regard the creation and appreciation of form by the human mind as an act of artistry, whether the artifact be a scientific theory, a machine, a sonata, a city plan, or a new design in human relations. And I believe and seek to show that in all these acts of artistry, intuition and rationality are always involved, usually in the roles of creator and critic.

In the next section I explore the basis for this dualism. We know something about the processes of perception, cognition, and recognition. We know that we come to recognize repetitions and regularities in the physical world long before we have any theories about why these should be. We know that unsupported toys fall from our cots before we know anything about the law of gravitation. We know the reliability or otherwise of mother's behavior long before we know any psychology. Our knowledge is contextual before it extends to causality; and it grows in both dimensions half-independently. We learn to distinguish more subtly differentiated contexts, just as we learn to distinguish the operations of more generalized laws. And equally, we learn to envisage and create new contexts, just as we learn to detect new causal relations. The technological innovator is a creator of new contexts, just as the scientific innovator is a discoverer of new causal regularities often based on his discovery, or even creation, of new conceptual entities such as elements or particles. Synthesis and analysis, contextual and causal explanation are distinct though inseparable aspects of human mental process in all mental activities. It need cause us no surprise that they are equally manifest in physics and mathematics.

I stress the tacit nature of the standards which we develop to guide our intuitive processes because this has become a stumbling block to the "rational" understanding of "intuition," an aspiration which is obviously not fully attainable if the two are complementary capacities of the human mind.

In a later section I examine a process of design where the imposition of form on experience is more conscious and more obvious. I seek to show that in this case also the choice between possible forms is not governed by criteria which are fully describable and for the same reason.

In a brief last section I summarize the epistemological conclusions to which these reflections lead. They may not yet be orthodox, but they are far more constant with the thinking of our time than they would have been even a decade ago.

According to the view put forward here, knowing and designing are not separate or even separable activities, since our whole schema for knowing is a design, a model of reality consciously and unconsciously made, and constantly revised. Moreover it is a selective model made in response to our concerns which alone determine what we regard as relevant enough to be worth modeling. The design produced by the natural and the logical sciences is more conditioned by independent variables which it cannot "redesign." But it is a design for all that and a design that is intimately connected with the concerns that drive us to make it, concerns that notably include aesthetic satisfaction.

Perception, Cognition, and Recognition

Professor Christopher Alexander, in his book *Notes on the Synthesis of Form*,[4] says, in effect, that design does not consist in the realization of form but in the elimination of "misfit." The designer approaches his task with a set of tacit criteria, which appear only when some specific design is found to be inconsistent with one of them. The norm is known only negatively, when it is infringed. For the state of "fit" we have no evidence, except the agreeable absence of misfit. We have scarcely even a vocabulary for it—how vague and how numerous are all the antitheses to pain! Alexander observes that this elusive quality of the norm has been noted in other fields also; he instances the difficulty experienced by doctors when they try to define "positive health," and by psychiatrists when they try to define psychological normality.[5]

I believe that Alexander's insight is of great generality and importance, and I shall develop it in ways which go beyond his statement and with which he might not agree. I shall postulate, as the basic fact in the organization of experience, the evolution of norms by which subsequent experience is ordered, and which are themselves developed by the activities that they mediate. I shall suggest that this evolution of norms is a fundamental form of learning; that it provides the criteria not only for ethics and aesthetics, but also for all forms of discrimination (including those used by the various sciences), and that the norms so developed are tacit by logical necessity.

Suppose, for example, that I say, "That is an ash tree." If you ask why I think so, I can only reply, "Because it looks like one." If you are not satisfied, we may approach the tree, examine its leaves and the character of its bark and its seeds, if it happens to be seeding. This analysis may or may not confirm my initial judgment but it played no part in making that judgment. The tree was too far away for me to see these details.

If I had said, "That is a beautiful ash tree," you would have regarded my judgment as "aesthetic" and we might have discussed the basis for my judgment that the tree was beautiful. Did I, for example, mean merely that is was an exceptionally fine specimen of its kind? It is less common to class as an exercise of aesthetic judgment the ability to classify it correctly as an ash tree, irrespective of my emotional response to it. This, nonetheless, is the wide sense in which I am using the words "aesthetic" and "intuitive." The recognition of form is an exercise of judgment made by reference to criteria which are not fully describable because of the subtle combination of relationships in which they reside and equally because of their dependence on *context*.

I use the word "norm" in an unusually wide sense, to cover the criterion for every judgment which classifies, whether it seems to involve a judgment of fact or of value. This distinction itself I regard as outmoded for more reasons than I shall have space to include. Professor Pitkin,[6] in her book *Wittgenstein and Justice*, has shown that the many different forms in which this antithesis is expressed (the "is" and the "ought," descriptive and normative, and so on) have very different meanings. I shall stress that the *concern* of a *human* mind is necessary to define any situation and perhaps necessary to define even what we call a fact, since a fact wholly irrelevant to any human concern would not be knowable.

The norms which are best understood scientifically are those which turn visual sensory input into perception. The child learns to recognize and to name, partly by being often exposed to the same stimuli, partly by its own inner activity of ordering its experience, and partly by the persuasion of other human beings, exhorting, encouraging, correcting. In some way not yet fully understood, his central nervous system develops readiness to group together, attend to, and recognize aspects of his surround—faces, places, belongings, relations—which recur and are of interest to him and to organize his accumulating knowledge by classifying it in an increasing number of overlapping categories.[7]

These "readinesses to classify" are commonly called schemata. The word "schema" is important for my purposes because it is the only accepted word in a class much wider than that in which it is commonly used. We clearly develop "readinesses to recognize" not only perceptual gestalten but also situations of great generality and complexity (such as illness and revolution) and concepts of great abstraction (such as entropy and the British constitution). We develop schemata, perceptual and conceptual, partly by being exposed to countless particular examples from which we abstract what they have in common for our purposes (as a doctor does in a hospital or a lawyer in the courts) and partly through the "open-endedness" of language, introducing us to abstractions which later examples make real (like a doctor with his textbook of physiology and a lawyer with his textbook of jurisprudence). It is commonly recognized that a combination of the two is the best way to develop those readinesses to recognize which it is the business of education to teach and of all ages to learn.

The duality of this process, though familiar to experience, has long been an offense to the Western scientific mind and has given rise to a long-drawn controversy whether the mind *identifies* the familiar by checking a list of characteristics which define its identity or *recognizes* it by fitting some kind of perceptual gestalt to some kind of mental template. Adherents of either view can find plenty of weaknesses in the other,[8] but neither party, until recently, seems to have conceived the possiblity that the brain might be capable of both processes and might use them in appropriate—or sometimes inappropriate—combinations. This, nonetheless, seems to be the fact. Brain scientists are much concerned with the neurological basis for this in the difference of function between the two hemispheres of the neocortex. I am not concerned with the problems of location which engage them, but I am intensely concerned with their finding that the human brain is indeed capable of what Dr. Galin[9] calls "two cognitive styles" of activity.

The child learns to see. So does the beneficiary of corneal grafting. So does the doctor learning to diagnose; the radiologist learning to read a radiograph; the stockbreeder and forester learning to distinguish a good specimen from a poor or sick one. So does the connoisseur of Chinese ceramics. All these people, later, can write books about the criteria they use, but they cannot express these fully in a rule which the inexperienced can apply. The future master must make these schemata his own by frequent use; and these schemata are also criteria, instruments by which specific misfits are detected, though they themselves cannot be specified.

Professor Woodger[10] instances the novice looking through a microscope for the first time. He has a visual experience, but he does not perceive anything because he has not yet built up the schemata by which to recognize the inhabitants of this elfin world. Cognition is the result, as well as the precursor, of *re*cognition. G.H. Lewes[11] expressed this elegantly and generally as long ago as 1879, ". . . the new object presented to sense or the new idea presented to thought must also be *soluble in old experience*, be *re*cognized as like them, otherwise it will be unperceived, uncomprehended." (italics added)

Woodger has no use for the distinction between percepts and concepts. A percept is a concept. The link with the primary world of sensory experience is always tenuous and selective. A rabbit to an anatomist is a different bundle of abstractions from a rabbit to a cook—even if it be the same rabbit. Bruner, Goodnow, and Austin[12] agree with him that in perception, no less than in the most abstract thinking, the categories we use are our own invention; the order which we discover is imposed by ourselves and validated by its practical convenience to ourselves.

This is not to say that there is no order to be discovered in the natural world; on the contrary, the confirmation of experiment by the scientist and the less rigorous confirmation of the ordinary man's experience is taken as evidence that the order devised by the mind bears some valid relation to the order inherent in the "real world out there." We have at least constructed in our heads a viable analogue. But its viability is measured not only by its conformity with other experience of our own. It must also be sufficiently shared to mediate communication with others. Radical innovations in thinking take time to percolate into other minds and until they have done so, they are impotent and precarious.

Further, the validity of our chosen "order" is measured also by its power to make our own experience acceptable to ourselves. For this it must be sufficiently concordant with the rest of our organizing concepts; and it must also create a world in which we can bear to live. Rokeach,[13] referring to "belief systems" (which correspond closely to what I have called "appreciative systems"), writes, "Such systems . . . serve two opposing sets of functions. On the one hand, they are Everyman's theory for understanding the world he lives in. On the other hand, they represent Everyman's defense network, through which information is filtered in order to render harmless that which threatens his ego . . . a belief system seems to be constructed to serve both masters at once; to understand the world in so far as possible and to defend against it in so far as necessary." This defensive function is not necessarily pathologic, though it always has a cost.

Thus, the world of reflective consciousness—I will call it the appreciated world—in which each of us lives, is structured by readinesses to conceive things and relations in particular ways, readinesses which are developed in our brains by experience, including experience received through human communication. I will extend the word "schemata" to cover all such readinesses, since it is free from the normative implications of such words as "standard," "criterion," and "norm" itself. Nonetheless, such schemata do function as norms, standards, and criteria, even in the most purely factual acts of discrimination. Screening experience, they classify what "fits" and reject "misfits." And they do so equally whether they define a state of affairs in my surround—"That is a bull," or its implications for me—"That is a threat," or a situation accepted by myself or others as requiring a particular response—"That is an obligation."[14]

Some Epistemological Implications

I labor these familiar points because I want to rescue from their normal oblivion three facts which I believe to be highly important: First, facts are not data. They are mental artifacts, selected by human concerns and abstracted from experience by filtering through a screen of schemata. Second, this screen is necessarily tacit; we infer its nature only from observing its operations, but our inferences can never be complete or up to date. Third, the screen is itself a product of the process which it mediates and, though tacit, can be developed by deliberately exposing it to what we want to influence it. (This is the essence of education.)

These schemata do not exist in isolation. They develop within the multiple contexts of experience. I find it convenient to think of these contexts as ordered by a three-dimensional matrix. What we notice is selected by our concerns, and our concerns are excited by what we notice. I will call our concerns our "value system" and call our organized readinesses to notice our "reality system."[15] I think of these as forming two sides of the matrix. The third is, of course, the dimension of time. Our reality system can represent the future and the hypothetical, as well as the actual present, and our value system can both evoke and respond to such constructions. Our most familiar mismatch signals are generated by the comparison of our expectations with our fears and our aspirations—that is to say, by comparing the constructions of our reality system and our value system when both are extended into the future.

I use and offer this simply as a convenient mental model. We are handicapped by lack of a realistic model of how our brains actually work, but communication science, by combining what we know of analogue and digital processes, can already provide us with a much more adequate picture than was possible even a few years ago, as Professor MacKay has shown.[16]

These schemata are systematically related; a change in one will involve some change in others and will be resisted in proportion to the extent of change involved unless this resistance is offset by the perceived benefit promised by the change. The theory of biological evolution, for example, when first put forward, was perceived by some as a hugely liberating idea, by others as hugely threatening. Hence the intense controversy which it aroused among laymen as well as scientists.

The impact of change will also be affected by the ease with which the proposed change can be understood. Biological evolution, however acceptable or unacceptable, was widely understandable, at least in principle. The theory of relativity was not. This difference muted resistance in some quarters and increased it in others.

T. S. Kuhn[17] in *The Structure of Scientific Revolutions* has drawn a distinction between the normal course by which scientific knowledge grows by accretion and the periodic crises which call for a new "paradigm." He has pointed out that minds attuned to the "normal" course seldom initiate the paradigm shift, though they unconsciously prepare the way for it. He has also observed that the same process is to be seen in art. So long as art worked within the paradigm of representation, its achievements were indeed cumulative. It progressively learned new ways to represent three-dimensional scenes on two-dimensional space. A new paradigm set new standards—not higher standards for the same kind of excellence but standards related to a different kind of excellence.

In fact, even what Kuhn calls "normal" science does not proceed without minor shifts in "tacit norms." The simplest piece of induction is not explicable or even describable in the way in which we can describe and demonstrate the most complex deductive process. Nonetheless, paradigm shifts are most dramatically visible in the development of individual scientists and individual artists when they occur suddenly and make a major difference. Consider, for example, the story of Kepler, transported with excitement at suddenly seeing in Tycho Brahe's calculations what Tycho himself could not see—that they were consistent only with elliptical planetary paths, a concept banned from consideration by the authority of Aristotle. Most dramatic of all perhaps is the story of Kekule, seeing in a dream or vision as he dozed before the fire the benzene ring in the form of a whirling fiery serpent eating its own tail.

Wertheimer[18] claims that his discussions with Einstein showed how critical was the moment when it first occurred to Einstein to question the conventional concept of time. And although Miller[19] has criticized Wertheimer's reconstruction of Einstein's thought processes, his own well-documented account in this book of those processes, not only in Einstein but in other leading minds who pioneered the amazing subsequent development of quantum theory, provides even more abundant examples of their paradigm shifts. It also reveals how closely these shifts are related to the personality and experience of the particular scientist involved. "Making sense" of the world is, it seems, a highly individual activity, even where the subject matter of our enquiry is the apparently independent fields of the "natural" and the "logical" sciences. Why else should Heisenberg, fully at home in a nonvisualizable universe, call Niels Bohr's mathematics "disgusting" as described by Professor Miller in his paper in this book?

Similar shifts are seen in the development of individual artists. The development of Picasso's art, as of many others, in other media as well as painting, illustrates the self-generated shifts of paradigm which a creative mind can achieve. The breakthrough comes sometimes after a spell of inactivity either willed or imposed by the artist's incapacity to produce. Sometimes it can be seen in retrospect as following a series of tentative struggles towards what, for the artist, was still a hidden goal. In either case, when it emerges it is unmistakable.

These dramatic shifts make visible with peculiar clarity the structure of tacit norms, previously taken for granted, which they assail and replace. I am equally concerned in this paper with the process by which a system of tacit norms changes gradually over time.

I will next examine an example of design of a different kind, the redesign of an urban environment. The effort is far more conscious. Nonetheless, the norms involved remain only half revealed both in the process by which the problem is, ultimately, defined and in the process by which one of many possible partial solutions is chosen. And the reciprocal effect of the effort on the norms by which it is guided, though not fully detectable even with the wisdom of hindsight, is no less important than in the example already examined.

Design as the Resolution of Conflict between Norms

Consider a problem of urban planning. Several criteria can be described in general terms. Buildings must have access to vehicular and pedestrian traffic appropriate to the activities which they generate; and the two types of traffic must be sufficiently separated to preserve an acceptable level of safety. Noise, air pollution, and interference with light must be kept within acceptable thresholds. And so on. But what in each case is the level of the appropriate and the acceptable? The policymaker may allot target values in each case but he can be sure that, if he pitches them high enough to be unquestioned, some at least, will not be capable of being realized. He cannot even make an exhaustive list, until an actual plan begins to emerge, so that its actual effect can be envisaged in each of the dimensions of success. He must wait for the planner before he can clearly define the problem that he wants the planner to solve.

What of the planner?—in so far as his function can be separated. Each of the requirements which he has to satisfy (and which thus become his concerns) make relevant, as possibilities or limitations, features of the physical site; and these in turn suggest their relevance, for good and ill, to other requirements. Hypothetical solutions begin to shape themselves in his mind and in rough sketch plans; and these engender often unsuspected meanings as they intersect with the various requirements in which he is concerned or even suggest others with which he ought to be concerned. One cell after another in the matrix is activated by such intersections of concern and opportunity; and from each intersection streams of further activation resonate along all the dimensions of the matrix.

This exercise can easily lose itself in boundless complexity. The list of requirements and the facts relevant to each can be extended in number and time with no clear limit, and every extension multiplies their interactions with each other. Alexander, in the book already mentioned,[20] makes some valuable suggestions for keeping these many-factored problems under control by identifying those variables which can be grouped together in relatively independent clusters. But in any but the simplest problem, it is, I believe, vain to hope for a solution which will produce for every requirement a "fit" which would have been regarded as acceptable when the exercise began. Any solution will have to deal with some requirements in a way which will become acceptable only in the light of what it will make possible in other dimensions of success or, alternatively, of what is then seen to be the cost of making it any better, when this cost is measured in terms of the limitations it would pose on satisfying other requirements.[21]

The designer, then, like the scientist, is engaged in a synthetic exercise. He must produce a single design which will be judged by multiple criteria. Some of these reinforce each other; some conflict with each other; most compete with each other for scarce resources. All are affected in some degree by any change made in the interest of one of them. The number of possible designs, even within given costs, is unlimited and unknowable, for it depends on possibilities of innovation which cannot be known before they have been made. The comparison of one with another can be made only when both have been worked out and even then depends on the relative value attached to disparate criteria within the framework of a single solution.

However great the number of possible designs, the number submitted to the policy maker is seldom more than one. The resources demanded by large-scale planning are too great to permit detailed alternatives. Hence enormous importance attaches to the rapid and often obscure process by which the basic lines of the proffered solution are chosen, for these soon generate many vested interests, valid as well as invalid. Not least of these is that its sponsors, having grown familiar with its implications, can more confidently exclude the possibliities of unwelcome surprise, which would lurk in any alternative, until it had reached the same degree of elaboration.

The successful designer chooses what proves to be a viable approach by a process which is much better than random and which seems sometimes to be guided by uncanny prescience. So does the technological inventor, the scientific discoverer, the successful policy maker in government and business, and those apparently ordinary mortals whose human relations are at once richer, more varied, and more orderly than those of their neighbors. I do not postulate any unknown mental function—or, at any rate, any more unknown than they all are—when I describe these gifted people as having (like the artist) unusual sensitivity to form. But in thus grouping them together, I do suggest that they have something in common in terms of cerebral organizing capacity. I have no doubt that this something is a specially happy combination of the "two cognitive styles" mentioned by Dr. Galin in the paper already referred to—the one logical, analytic, and explicit; the other (and, in these cases, the more important) contextual, synthetic, and tacit.

Giftedness in Rationality and Intuition

This tentative postulate seems to derive support from many sides. The capacity for sensory discrimination varies greatly between individuals. It is most easily charted in music because the sound patterns produced by musical instruments can be formally described,[22] even when they are immensely complex, as in concerted orchestral passages. It is demonstrable that people differ, not merely through differences in training, in their ability to recognize, for example, variations on a theme. It seems reasonable to suppose that this innate capacity for discriminating musical patterns sets limits both to musical interest and to musical achievement.

By discrimination in its most general sense I mean the ability to distinguish figure from ground, signal from noise.[23] It is the basic limitation of any information system. It is distributed between individuals not only unevenly but selectively. Those concerned with the study of gifted children distinguish at least four kinds of giftedness, each of which can be described as unusual power of discrimination. The most familiar are the intellectually gifted, who can be identified with some confidence by tests of ability to recognize logical, including mathematical, relations. Distinct from these are the inventive, whose ability for practical innovation can coexist with quite limited power to handle abstractions. The aesthetically gifted are again a class apart. They are often impractical and sometimes unintellectual; and their gift for appreciating sensory form is highly selective. Aesthetic appreciation of nature may be dissociated from the appreciation of the fine and applied arts; and within these last, sensitivity to one medium may not extend to others.

In the categories of conscious experience, these forms of giftedness cover a wide variety of gifts, but they are all mediated by the brain and central nervous system. They all involve discrimination, in some form, between figure and ground, signal and noise; and they all depend on tacit criteria developed by experience within the inherent limitations of the particular neural heritage.

Some students of human giftedness distinguish a fourth type—social giftedness. These are those who show unusual interest and ability in sensing, maintaining, and creating relations with other people. These gifts too would seem to depend on unusually high capacity for discrimination. Students of human dialogue can show that it involves each party in setting up an inner representation of the other, and that the level of dialogue depends not only on the accuracy and refinement of this model, but also on the attitude of each to his inner representation of the other. G.H. Mead insisted on the social importance of a variable which he described as the ability and willingness to take the generalized role of the other. Communication theory begins to make this concept more precise.

I have distinguished two functions which in practice are never wholly separated but which are, nonetheless, logically distinct as two reciprocating phases in a recurrent process of mental activity. One is the creative process, which presents for judgment a work responsive to many explicit and tacit criteria. The other is the appreciative process, which judges the work by the criteria, tacit as well as explicit, to which it appeals, and finds it good or wanting, better or worse than another. The two phases of the process may alternate many times in the course of producing the work. The work may never be finished; in a sense it can never be finished, for it is part of an ongoing process. Even the individual works of an artist are part of his "work," which ends only with his death, or when he has nothing more to say, and which continues even after that in the creative and appreciative minds which it quickens.

In the example last given, the form of the work was an urban design. An urban design may be viewed as a creation in any of the four fields in which we exercise judgment—scientific, technological, ethical, and aesthetic. It may be viewed as a work of art, appealing to aesthetic criteria, like a sculpture or a painting. It may be viewed as an invention, appealing to functional criteria of utility. It may be viewed as a social creation, appealing to criteria of social need and satisfaction. It may be viewed as an intellectual creation, an expression of abstract relations, like a scientific theory. Examples could be chosen which would more clearly emphasise any one of these aspects, rather than another. We may be right to distinguish sharply between these different kinds of knowing and their related criteria. Yet they have notable common features which are likely to correspond to common features in the working of the human brain or in the patterns which our culture imposes on it.

The Dynamics of Change in Normative Systems

The common features I want to emphasize are the following:

1. The form is produced by the activity of a concerned mind structured by tacit norms as well as by explicit rules. This concerned mind abstracts for attention what I will call a *situation*, by which I mean a set of related facts relevant to its concern. Part of this situation is seen as not modifiable by the agent; I will call this the *context*. The rest of the situation is the area to which form is to be given. I will call it the *field*.[24]

2. The form is specific. It is to be realized in particular terms, by arranging the field in particular ways, in relation to the context.

3. Both phases of the process by which form is given to the field change the norms to which they consciously and unconsciously appeal. The appreciative phase changes them by the mere fact of using them to analyze and evaluate a concrete situation, for this may affect both their cognitive and their evaluative settings. The creative phase affects them by presenting new hypothetical forms for appreciation. The realization of the chosen form still further affects the norms involved, for it affects the situation, including its division between field and context. It would thus alter the stream of match and mismatch signals generated by the situation, even if it had not already altered the setting of the norms themselves.

This process of change is the focus of my attention, for I believe it is the key to our understanding of our predicament and of the scope of our initiative. I have argued that we know anything at all only by virtue of a system of largely tacit norms, developed by individual and social experience, which itself is structured by our individual concerns, and that this system has the threefold task of guiding action, mediating communication, and making personal experience meaningful and tolerable. It can change only at a limited rate, if it is not to fail in one or more of its functions. Its failures at the level of the individual can be studied in any mental hospital, and at the social level, in all the more disturbed periods of history, notably the present. Hence the importance of understanding the process of change, its possible patterns, and its inherent possibilities and limitations.

The system of tacit norms, which I call an appreciative system, tends to be self-perpetuating. Our mutual understanding and cooperation, our powers of prediction and effective action depend on its being widely shared and accepted. So any challenge to it awakens protective responses. Each generation has a powerful interest in transmitting it to the next. Representing as it does the accumulation of experience, it is supported both by authority and history.

In the social field it is also to some extent self-validating, since sanctioned mutual expectations tend to elicit the behavior which will confirm them. In the field of the natural sciences, where the variables are more independent, this conservatism is less likely to be self-validating though it may long inhibit change. Michael Polanyi[25] lived long enough to see the adoption in his lifetime of a scientific hypothesis formulated by him forty years before, but barred from acceptance in the meantime by its departure from the then most acceptable style of explanation. In his account of this experience he expresses his approval of this degree of inertia, even though it nearly cost him his scientific career.

On the other hand, such systems also contain within themselves the seeds of their own reversals. Each is a work of art, however unconscious, and, like all works of art, attains form only by a process of selection which excludes possible alternative forms. These in time clamour for realization. They are kept alive in the meantime in those individuals and subcultures which are least satisfied by the accepted systems; and they grow at the expense of the accepted system as soon as that system ceases to command the confidence and authority of its heyday.

Furthermore and more conspicuously, the accepted system is challenged by changes in the context, often brought about indirectly by its own development which renders that context no longer appropriate. These changes may be in the physical or the institutional or the social or even the intellectual context. All are abundantly illustrated in the recent history of the Western world. Physical exploitation has posed problems of pollution which turn growth from a promise into a threat. Market institutions, developing, have changed the nature of the market. Democratic political institutions, developing, have transformed the concept of democracy. "Liberal" values have made a world which increasingly rejects liberal values. Styles of scientific thinking, pushed to their extremes, reveal their limitations and subsume or are overwhelmed by their rivals. "Teleology," for example, a word wholly unacceptable to science even fifty years ago, attained respectability almost overnight as soon as it could be applied to manmade *machines*.

Finally, the accepted system may be challenged by collision with rival appreciative systems. This also is being illustrated by contemporary history as never before. In a world where interactions multiply on a planetary scale, inconsistent subcultures multiply by fission to attest the passionate need of each individual for an apt and shared appreciative system, however small the sharing group. It is in no way surprising that increased physical contacts across the world should have called into being not "one world" but more mutually antagonistic worlds than ever before.

It may be objected that this, even though true in those fields of *scientia* which are affected by human culture, is not true of the natural sciences. The world's atomic scientists talk a common language, even though the world's politicians do not. I have argued that this is true only as a matter of degree.

Conclusions

It seems to me possible, in the light of these ideas, to arrange the different fields of potential knowledge in an order which explains both the extent to which they are open to human knowing and the extent to which our acceptance of our knowledge rests on its survival of rational tests on the one hand and its congruence with tacit standards of form on the other. This "order" does not involve sharp breaks between "scientific knowledge" and "unsupported beliefs."

It is generally recognized today that even in the natural sciences the scientific method does not "validate" its hypotheses. It can only test them and attach increased credence as they survive those tests. It is recognized also that credence develops at least as much from the congruence of the theory with the existing body of knowledge and from its facility in explaining facts other than those which it was devised to explain and in generating further hypotheses which depend on it.

Less commonly recognized is the limited extent of the knowledge confirmed even to this extent and the status of the remainder. Even in the natural sciences theories rightly retain their power even though well established facts show that they must be at least incomplete. For example, the ascertained facts of what is now (perhaps mistakenly) called extrasensory perception show that our current ideas of sensory perception must be at least incomplete. But in the absence of a theory to link what we do not understand with what we do understand, these facts await incorporation in the general body of knowledge, just as the evidence for biological evolution had to await its explanatory theory. Even an adequate theory may have to wait long for acceptance, as Polanyi's experience showed, for no better reason than its departure from current fashions of explanation.

Equally often rival theories compete to explain the same set of facts, as did Ptolemaic and Copernican astronomy. The judgment that accepted the second was not the result of "rational" testing. It was not even a preference for simplicity. For Copernicus, retaining circular planetary motion, had as much mathematical difficulty in accounting for his observations as did his Ptolemaic predecessors. Yet his theory, rightly in my view, bears his name, rather than that of Kepler who first gave the theory its manifest superiority in simplicity.

The situation is even more extreme in the field which Herbert Simon has called the "Artificial." For example, it is sometimes objected that most psychoanalytic theory is "unscientific" because it cannot be *dis*proved. This may be so, but we are not therefore irrational in accepting it in so far as we judge it to serve our need better than others or better than none. Human motivation is complex and culture-bound. Why should we expect to "understand" it completely or once and for all?

The example is also a disturbing reminder of the relation between knowledge and design in the field of the "artificial." However unsupported these theories may be deemed to be, countless families in the past seventy years have made them *true for them* simply by accepting them. And more generally they have become part of Western culture to the extent that they have affected our basic assumptions about the areas in which they operate.

It is a minor example of this, for Western culture has been profoundly affected by the findings, the methodology, the attitudes, and the outlook of scientists. It is not wholly the fault of scientists that what has passed into the general culture is grossly distorted in two critical ways. One is the mistaken identification of science with rationality. The other is the exaggerated dichotomy between science and nonscience. These two errors are nonetheless a grevious threat to our understanding of our own mental processes, for they ignore two basic facts of human epistemology.

The first of these is that our basic knowledge of the world, our neighbors, and ourselves is a set of expectations based on exposure to the regularities of experience. Science has vastly amplified and refined these expectations, but what lies outside its reach is by far the greater part.

Second, the appreciation of form based on tacit standards is as basic to science as to the much wider area of our tacit and explicit assumptions. It emerges most clearly from a study of the great innovations in science, but it is equally important and far more common in conserving and securing general acceptance for the common assumptions on which all our cooperative activity proceeds, not least the cooperative activity of science.

Notes and References

1. Herbert Simon, *The Sciences of the Artificial* (Cambridge: The MIT Press, 1969).

2. A. O. Lovejoy, *The Great Chain of Being* (Cambridge: Harvard University Press, 1966).

3. In fact I understand that the theory of particulate inheritance has itself been so qualified by deeper understanding of the interactions of the gene pool that genetic theory today is more similar in its actual effect to that presumed in Darwin's day than the theory which emerged in 1905.

4. Christopher Alexander, *Notes on the Synthesis of Form* (Cambridge: Harvard University Press, 1967).

5. Alexander, *Synthesis of Form*, p. 198. note 21.

6. Hanna F. Pitkin, *Wittgenstein and Justice* (Berkeley: University of California Press, 1972).

7. The evidence on the subject of perception is conveniently summarized in M. L. Johnson Abercrombie, *The Anatomy of Judgment* (London: Hutchinson, 1960). The relation of cognitive capacity to neural development is of course the focus of most of the work of Piaget.

8. This was the focus of contention especially in the second decade of this century between gestalt psychologists (for example, Koehler, Wertheimer) and holistic philosophers (for example, Bergson, Smuts) on the one hand and traditional, analytic, reductionist science on the other. As so often occurs, the manifest facts on which the innovators were insisting were ignored because they were offered, or at least construed, as an "either-or" choice in which acceptance meant the rejection of equally valued insights on the other side. The debate continues to be bedevilled by the "either-or" disease even today when physics has blessed the concept of complementarity . Let us hope that neurophysiologists will prevail on the field on which psychologists and philosophers battled almost in vain.

9. David Galin, "Implications for Psychiatry of Left and Right Cerebral Specialisation," *Archives of General Psychiatry* , October 1974, vol. 31.

This paper also contains an extensive review of the literature on this subject. For a special study of its implications for pattern recognition, see Roland Puccetti, "Pattern Recognition in Computers and the Human Brain," *Brit. J. Phil. Sci.* 25 (1974):137–154.

10. Woodger, *Biological Principles* (London and New York: Harcourt, Brace, 1929).

11. G. H. Lewes, *Problems of Life and Mind.*
Quoted in M. L. Johnson Abercrombie's *Anatomy of Judgment.*

12. J. S. Bruner, J. J. Goodnow, and G. A. Austin, *A Study of Thinking* (New York: Wiley, 1956).

13. Milton Rokeach, *The Open and Closed Mind* (New York: Basic Books, 1960).

14. Professor Rhinelander, in his book *Is Man Incomprehensible to Man*, expresses his opinion that this view is "essentially accurate" even though it "is at odds with much current philosophical theory and . . . bristles with controversial assertions and implications." (Portable Stanford, 1973, pp. 77, 78.)

15. I have developed this concept of interacting reality and value systems elsewhere, notably in *The Art of Judgment* (Londón: Chapman & Hall, and New York: Basic Books, 1965) ch. 4 and *Value Systems and Social Process* (London: Tavistock Publications and New York: Basic Books, 1968), ch. 9.

16. Notably in a paper "Digits and Analogues," published in the proceedings of the 1968 AGARD Bionics Symposium, Brussels, which also contains the major references to his earlier works.

17. T. S. Kuhn, *The Structure of Scientific Revolutions* (Chicago: University of Chicago Press, 1970).

18. See M. Wertheimer, *Productive Thinking* (New York: Harpers, 1959) p. 214.

19. Arthur I. Miller, "Albert Einstein and Max Wertheimer: A Gestalt Psychologist View of the Genesis of Special Relativity Theory," *Hist. Sci.* XIII (1975): 75–103.

20. Alexander, *Synthesis of Form*, ch. 5.

21. Well illustrated in the *Buchanan Report on Traffic in Towns* (London: H.M.S.O., 1963, p. 16)

22. Jeanne Bamberger distinguishes the formal structure of music from the (much less describable) figural structure imposed by the performer and the hearer and has shown that young children impose a figural pattern even on a sound sequence from which the performer has eliminated all but formal elements. ("The Development of Musical Intelligence I," July 1975, unpublished.) She is Associate Professor of Education, Division for Study and Research in Education, Massachusetts Institute of Technology and Associate Professor of Humanities (Music) at the same institution.

23. When I refer to "distinguishing signal from noise" I do not wish to imply that there is necessarily only one distinction to be made. Many alternative divisions of figure and ground may be possible. Dr. Hans Selye has described how, as a young medical student making his first contact

with hospital wards, he was struck not by the variety of the patients' symptoms but by the similarity which distinguished them all from the nurses and doctors around them. They *all looked ill*. He learned his appointed lesson, which was to distinguish and diagnose their disease. But he did not forget his initial insight. It was later to inspire what was to be his predominant life work as a researcher—the study of the body's response to stress of any kind—which he was to call the general adaptation syndrome. Hans Selye, *The Stress of Life* (New York: McGraw-Hill, 1956) pp. 14–17.

24. I take this use of the word "context" from Alexander (4).

25. Michael Polanyi, "The Potential Theory of Adsorption," *Science* 141, no. 3585, pp. 1010–1013.

SELECTED BIBLIOGRAPHY

Abercrombie, M. L. Johnson. *The Anatomy of Judgment*. London: Hutchinson, 1960.

Achinstein, Peter. "Models, Analogies, and Theories." *Philosophy of Science*, vol. 31, 1964: 328–350.

Arnheim, Rudolf, ed. *Entropy and Art*. Berkeley: University of California Press, 1971.

———. *Visual Thinking*. Berkeley: University of California Press, 1969.

Benthall, Jonathan. *Science and Technology in Art Today*. London: Thames and Hudson, Ltd., 1972.

Bohr, Niels. *Essays 1958–1962, On Atomic Physics and Human Knowledge*. New York: Interscience Publishers, 1963.

Born, Max. *Natural Philosophy of Cause and Chance*. Oxford: At the Clarendon Press, 1948.

Bronowski, J. *The Common Sense of Science*. New York: Vintage Books, n.d.

Bruner, Jerome S. *On Knowing: Essays for the Left Hand*. Cambridge: Harvard Univ.Press, 1962.

Bruner, Goodnow, and Austin. *A Study of Thinking*. New York: Wiley, 1956.

Buchanan, Scott. *Poetry and Mathematics*. New York: Lippincott, 1962. (first published in 1929)

Burke, John G. *Origins of the Science of Crystals*. Berkeley: University of California Press, 1966.

Cassirer, Ernst. *The Problem of Knowledge*. New Haven: Yale University Press, 1950.

Cohn, Myron A., ed. *Essays on Creativity in the Sciences*. N.Y.U. Creative Science Program Seminars, 1963.

Critchlow, Keith. *Order in Space*. London: Thames and Hudson; New York: Viking Press, 1969.

Daedalus: Creativity and Learning, vol. 94, no. 3, Summer 1965.

Daedalus: The Making of Modern Science: Biographical Studies, vol. 99, no. 4, Fall 1970.

Danzig, Tobias. *Number: the Language of Science*. New York: Macmillan, 1954.

Dirac, P. A. M. "Evolution of the Physicist's Picture of Nature." *Scientific American*, May 1963: 45–53.

Einstein, Albert and Infeld, Leopold. *The Evolution of Physics*. New York: Simon and Schuster, 1960.

Frank, Philipp. "A Colloquy on the Unity of Learning." *Daedalus*, vol. 87, no. 4, Fall 1958: 158–160.

———. "Contemporary Science and the Contemporary World View." *Daedalus*, Vol. 87, no. 1, Winter 1958: 57–66.

Gingerich, Owen, ed. *The Nature of Scientific Discovery*. Washington, DC: Smithsonian Institute Press, 1975.

Gombrich, Ernst. *Art and Illusion*. New York: Pantheon, 1960; Princeton: Princeton University Press, 1972.

Hardy, Godfrey H. *A Mathematician's Apology*. rev. ed. Cambridge: At the University Press, 1969.

Heisenberg, Werner. *Across the Frontiers*. New York: Harper & Row, 1973.

———.*Physics and Beyond*. New York: Harper & Row, 1971.

———. *Physics and Philosophy*. New York: Harper and Row, 1958.

Hesse, Mary. *Models and Analogies in Science*. London: Sheed and Ward, 1963.

Holt, Michael. *Mathematics in Art*. London: Studia Vista; New York: Van Nostrand Reinhold, 1971.

Holton, Gerald, ed. *Science and Culture: A Study of Cohesive and Disjunctive Forces*. Boston: Houghton Mifflin, 1965.

———. *Thematic Origins of Scientific Thought: Kepler to Einstein*. Cambridge: Harvard University Press, 1973.

Hutten, E. H. "The Role of Models in Physics." *British Journal for the Philosophy of Science*, vol. 4, 1953–54: 284–301.

Ivins, William. *Art and Geometry*. New York: Dover Publications, 1946, 1964.

Kepes, Gyorgy. *Education of Vision*. New York: Braziller, 1965.

———. *Module, Proportion, Symmetry, Rhythm*. New York: Braziller, 1966.

———. *New Landscape in Art and Science*. Chicago: P. Theobald, 1956.

———. *Structure in Art and Science*. New York: Braziller, 1965.

Kepler, Johannes. *The Six-Cornered Snowflake*. Edited and translated by Colin Hardie. Oxford: Clarendon Press, 1966.

Kohler, Wolfgang. *Gestalt Psychology*. New York: Liveright, 1947.

Kubler, George. *The Shape of Time*. New Haven: Yale University Press, 1962.

Kuhn, Thomas. *The Copernican Revolution*. Cambridge: Harvard University Press, 1957.

———. *The Structure of Scientific Revolutions*. Chicago: University of Chicago Press, 1970.

Langer, Susanne. *Feeling and Form*. New York: Scribner, 1953.

———. *Mind: An Essay on Human Feeling*. Baltimore: Johns Hopkins University Press, 1967.

———. *Philosophy in a New Key: A Study in the Symbolism of Reason, Rite and Art*. Cambridge: Harvard University Press, 1942, 1957.

Lehmann, Otto. *Molekularphysik, mit besonderen Berüchtsichtigung mikroscopischen Untersuchungen und Anleitung zu solchen*, 2 vols., Leipzig, 1888–89.

Loeb, Arthur. *Space Structures: Their Harmony and Counterpoint*. Reading, Mass.: Addison-Wesley, 1976.

Macquet, Jacques. *Introduction to Aesthetic Anthropology*. Reading, Mass.: Addison-Wesley, 1971.

Morawski, Stefan. *Inquiries into the Fundamentals of Aesthetics*. Cambridge: The MIT Press, 1974.

Mueller, R. E. *The Science of Arts*. New York: John Day Co., 1967.

Mumford, Lewis. *Technics and Civilization*. New York: Harcourt, Brace and Co., 1963.

Oppenheimer, Robert. "The Growth of Science and the Structure of Culture." *Daedalus*, vol. 87, no. 1, Winter 1958: 67–76.

Pattee, Howard H., ed. *Hierarchy Theory: The Challange of Complex Systems*. New York: Braziller, 1973.

Penrose, Roger. "Role of Aesthetics in Pure and Applied Mathematical Research." *Bulletin of the Institute of Mathematics and Its Applications*. vol. 10, 1974: 266–271.

Pirsig, Robert. *Zen and the Art of Motorcycle Maintenance*. New York: Bantam Books, 1974.

Poincaré, Henri. *Science and Hypothesis*. New York: Dover Publications, 1952. (originally published in 1903)

———. *The Value of Science*. Translated by G. B. Halstead. New York: Dover Publications, 1958. (originally published in 1905)

———. *Science and Method*. Translated by F. Maitland. New York: Dover Publications, n.d. (originally published 1908)

Polanyi, Michael. *Personal Knowledge*. Chicago: University of Chicago Press, 1958.

———. *The Tacit Dimension*. New York: Doubleday, 1966.

Popper, Karl. *Conjectures and Refutations: The Growth of Scientific Knowledge*. New York: Basic Books, 1965.

———. *The Logic of Scientific Discovery*. New York: Basic Books, 1959.

Rosenblueth, Arturo and Wiener, Norbert. "Roles of Models in Science." *Philosophy of Science*, vol. XII, 1945: 316–321.

Santillana, Georgio de. *Reflections on Men and Ideas*. Cambridge: The MIT Press, 1968.

———. "The Role of Art in the Scientific Renaissance." *Critical Problems in the History of Science*. Madison: University of Wisconsin Press, 1959.

Saxl, F. *Lectures*. London: Warburg Institute, University of London, 1957.

Schmidt, Georg. *Kunst und Naturform: Form in Art and Nature*. Basel, Switz.: Basilius Press, 1960.

Simon, Herbert A. "The Architecture of Complexity." *Proceedings of the American Philosophical Society*, 106 (1962): 467–482.

Sivin, Nathan. *Chinese Science*. Cambridge: The MIT Press, 1973.

Smith, Cyril. "Art, Technology and Science: Notes on their Historical Interaction." *Technology and Culture*, October 1970: 493–549.

———. "A Highly Personal View of Science and Its History." *Annals of Science*, January 1977: 49–56.

———. *A History of Metallography*. Chicago: University of Chicago Press, 1960.

———. "Metallurgical Footnotes to the History of Art." *Proceedings of the American Philosophical Society* 116 (1972): 97–135.

———. "Metallurgy as a Human Experience." *Metallurgical Transactions*, 6A (1975): 603–623.

———. "On Art, Invention, and Technology." *Technology Review*, vol. 78, no. 7 (1976): 2–7.

———." Reflections on Technology and the Decorative Arts in the Nineteenth Century." In *Technological Innovation and the Decorative Arts*, edited by Quimby, L.A.M. and Earls, P.G., pp. 1–62. Charlottesville: The University Press of Virginia, 1974.

Stent, Gunther S. *The Coming of the Golden Age*. New York: The Natural History Press, 1969.

———. "Prematurity and Uniqueness in Scientific Discovery." *Scientific American*, December 1972: 84–93.

Stevens, Peter. *Patterns in Nature*. Boston: Little, Brown, 1974.

Taylor, A. M. *Imagination and the Growth of Science*. New York: Schocken Books, 1966, 1970.

Thomas, Lewis. *The Lives of a Cell*. New York: Viking Press, 1974.

Thompson, Sir D'Arcy. *On Growth and Form*. Cambridge: At the University Press, 1917, 1961.

Vickers, Geoffrey. *The Art of Judgment*. New York: Basic Books, 1965.

Von Mises, Richard. *Positivism, A Study in Human Understanding*. Cambridge: Harvard Univ. Press, 1951.

Waddington, C. H. *Behind Appearance*. Cambridge: The MIT Press, 1970.

Weiss, Paul. *Life, Order and Understanding*. Austin: University of Texas Press, 1970.

Weisskopf, Victor. *Knowledge and Wonder*. New York: Doubleday/Anchor Books, 1966.

———. *Physics in the Twentieth Century: Selected Essays*. Cambridge: The MIT Press, 1972.

Welsh, Alexander. "Theories of Science and Romance: 1870–1920." *Victorian Studies*, December 1973: 135–154.

Wertheimer, M. *Productive Thinking*. New York: Harpers, 1959.

Weyl, Hermann. *Symmetry*. Princeton: Princeton University Press, 1952.

Whitehead, Alfred North. *Science and Philosophy*. New York: Philosophical Library, 1948.

Whyte, Lancelot L. *Accent on Form*. New York: Harper & Row, 1954.

———. *Aspects of Form: A Symposium on Form in Nature and Art*. London: Humphries, 1951, 1968.

———, Wilson, and Wilson. *Hierarchical Structures*. New York: American Elsevier, 1969.

Wilson, R. R. "Physics and the Human Spirit." *Oppenheimer Memorial Lecture*, November 22, 1976.

INDEX